Selenium WebDriver Quick Start Guide

Write clear, readable, and reliable tests with Selenium WebDriver 3

Pinakin Chaubal

BIRMINGHAM - MUMBAI

Selenium WebDriver Quick Start Guide

Copyright © 2018 Packt Publishing

Commissioning Editor: Kunal Chaudari
Acquisition Editor: Siddharth Mandal
Content Development Editor: Smit Carvalho
Technical Editor: Sushmeeta Jena
Copy Editor: Safis Editing
Project Coordinator: Hardik Bhinde
Proofreader: Safis Editing
Indexer: Mariammal Chettiyar
Graphics: Alishon Mendonsa
Production Coordinator: Aparna Bhagat

First published: October 2018

Production reference: 2051118

Published by Packt Publishing Ltd.
Livery Place
35 Livery Street
Birmingham
B3 2PB, UK.

ISBN 978-1-78961-248-6

www.packtpub.com

`mapt.io`

Mapt is an online digital library that gives you full access to over 5,000 books and videos, as well as industry leading tools to help you plan your personal development and advance your career. For more information, please visit our website.

Why subscribe?

- Spend less time learning and more time coding with practical eBooks and Videos from over 4,000 industry professionals

- Improve your learning with Skill Plans built especially for you

- Get a free eBook or video every month

- Mapt is fully searchable

- Copy and paste, print, and bookmark content

Packt.com

Did you know that Packt offers eBook versions of every book published, with PDF and ePub files available? You can upgrade to the eBook version at `www.Packt.com` and as a print book customer, you are entitled to a discount on the eBook copy. Get in touch with us at `service@packtpub.com` for more details.

At `www.Packt.com`, you can also read a collection of free technical articles, sign up for a range of free newsletters, and receive exclusive discounts and offers on Packt books and eBooks.

Contributors

About the author

Pinakin Chaubal is a BE (Computer Science) with more than 18 years of experience in the IT industry. He is a PMP-certified professional and has worked with employers such as Patni, Accenture, L&T Infotech, and Polaris. He is currently working as a automation architect at Intellect Design Arena Ltd. (the product wing of Polaris). He has designed several frameworks using various techniques, including hybrid, keyword-driven, Page Object Model, and BDD, with Cucumber and Java. He has written one independently published book on Page Object Model using Selenium WebDriver and Java. He has been a reviewer for two books published by Packt. He has his own YouTube channel called *Automation Geek*, which covers various concepts related to testing and automation.

About the reviewer

Nilesh Kulkarni is a staff software engineer, currently at PayPal. Nilesh has extensive experience of working with Selenium. Nilesh has developed frameworks on top of WebDriver in different programming languages and is an open source contributor. Nilesh has actively worked on PayPal's open source UI automation framework, nemo.js. Nilesh is passionate about quality and has worked on different developer productivity tools. He often hangs out on Stack Overflow.

Packt is searching for authors like you

If you're interested in becoming an author for Packt, please visit authors.packtpub.com and apply today. We have worked with thousands of developers and tech professionals, just like you, to help them share their insight with the global tech community. You can make a general application, apply for a specific hot topic that we are recruiting an author for, or submit your own idea.

Table of Contents

Preface

Selenium WebDriver is based on the JSON wire protocol. This book explores various facets of Selenium WebDriver 3. It introduces Selenium WebDriver 3 in a layman fashion and opens the areas in Browser Automation to the reader.

Starting from a very basic introduction to element locators, the basic Selenium commands are explored and various programs are demonstrated to make the concepts clear. Handling popup windows and alerts is dealt with next, followed by various waiting mechanisms. Then we move on to the `Actions class` and JavaScript executor. Eventually, we explore the command design pattern, create few components of a Keyword-Driven framework, and learn about the extra locators available in Selenium WebDriver 3.

Who this book is for

This book is intended for people who wish to learn Selenium WebDriver from scratch. It can also be used by people working on other automation tools, such as UFT, and want to explore Selenium from the ground up.

What this book covers

`Chapter 1`, *Introducing Selenium WebDriver and Environment Setup*, gently introduces the reader to what Selenium is, how WebDriver is different from Selenium RC, and covers how to set up Eclipse.

`Chapter 2`, *Understanding the Document Object Model and Creating Customized XPaths*. covers with locator identifying mechanisms and the different ways to find XPath. It also introduces the Fillo API and debugging in Eclipse.

`Chapter 3`, *Basic Selenium Commands and Their Usage in Building a Framework*, covers the various Selenium commands and their practical usage. We also see some wrapper methods, which can be useful in designing a framework. We go over extract programs to fetch data from excel based on certain criteria.

`Chapter 4`, *Handling Popups, Frames, and Alerts*, covers how to handle modal and non-modal popups. We create some customized HTML pages with JavaScript for this purpose.

`Chapter 5`, *Synchronization*, covers the various ways of waiting for page loads, elements to be visible, and jQuery execution to get completed.

Chapter 6, *The Actions Class and JavascriptExecutor*, takes a look at the what the Actions class and the JavaScript executor are by going through many examples with the examples of HTML pages that we create.

Chapter 7, *The Command Pattern and Creating Components*, covers the command pattern and explains how we can use it to create a keyword-driven framework. We also learn about Selenium Grid.

Chapter 8, *Hybrid Framework*, explores TestNG listeners and the WebDriverManager library. We also learn how to create a report from TestNG.xml.

To get the most out of this book

- You will need an understanding of OOPs concepts in Java
- Start from the first chapter and move chapter by chapter until you feel confident
- Keep creating tests and sample pages on which you can test
- Work on the exercises in the book.

Download the example code files

You can download the example code files for this book from your account at www.Packt.com. If you purchased this book elsewhere, you can visit www.Packt.com/support and register to have the files emailed directly to you.

You can download the code files by following these steps:

1. Log in or register at www.Packt.com.
2. Select the **SUPPORT** tab.
3. Click on **Code Downloads & Errata**.
4. Enter the name of the book in the **Search** box and follow the onscreen instructions.

Once the file is downloaded, please make sure that you unzip or extract the folder using the latest version of:

- WinRAR/7-Zip for Windows
- Zipeg/iZip/UnRarX for Mac
- 7-Zip/PeaZip for Linux

The code bundle for the book is also hosted on GitHub at `https://github.com/PacktPublishing/Selenium-WebDriver-Quick-Start-Guide`. In case there's an update to the code, it will be updated on the existing GitHub repository.

We also have other code bundles from our rich catalog of books and videos available at `https://github.com/PacktPublishing/`. Check them out!

Download the color images

We also provide a PDF file that has color images of the screenshots/diagrams used in this book. You can download it here: `https://www.packtpub.com/sites/default/files/downloads/9781789612486_ColorImages.pdf`.

Code in action

Visit the following link to check out videos of the code being run:
`http://bit.ly/2CGHvf6`

Conventions used

There are a number of text conventions used throughout this book.

`CodeInText`: Indicates code words in text, database table names, folder names, filenames, file extensions, pathnames, dummy URLs, user input, and Twitter handles. Here is an example: "Mount the downloaded `WebStorm-10*.dmg` disk image file as another disk in your system."

A block of code is set as follows:

```
<dependency>
    <groupId>com.codoid.products</groupId>
    <artifactId>fillo</artifactId>
    <version>1.18</version>
</dependency>
```

When we wish to draw your attention to a particular part of a code block, the relevant lines or items are set in bold:

```
This piece of code will print out
TestCase1
TestCase3
TestCase4
```

Any command-line input or output is written as follows:

```
$ mkdir css
$ cd css
```

Bold: Indicates a new term, an important word, or words that you see onscreen. For example, words in menus or dialog boxes appear in the text like this. Here is an example: "Select **System info** from the **Administration** panel."

 Warnings or important notes appear like this.

 Tips and tricks appear like this.

Get in touch

Feedback from our readers is always welcome.

General feedback: If you have questions about any aspect of this book, mention the book title in the subject of your message and email us at customercare@packtpub.com.

Errata: Although we have taken every care to ensure the accuracy of our content, mistakes do happen. If you have found a mistake in this book, we would be grateful if you would report this to us. Please visit www.packt.com/submit-errata, selecting your book, clicking on the Errata Submission Form link, and entering the details.

Piracy: If you come across any illegal copies of our works in any form on the Internet, we would be grateful if you would provide us with the location address or website name. Please contact us at copyright@packt.com with a link to the material.

If you are interested in becoming an author: If there is a topic that you have expertise in and you are interested in either writing or contributing to a book, please visit authors.packtpub.com.

Reviews

Please leave a review. Once you have read and used this book, why not leave a review on the site that you purchased it from? Potential readers can then see and use your unbiased opinion to make purchase decisions, we at Packt can understand what you think about our products, and our authors can see your feedback on their book. Thank you!

For more information about Packt, please visit `packt.com`.

1
Introducing Selenium WebDriver and Environment Setup

Welcome to the exciting world of test automation using Java 8 and Selenium WebDriver 3.x. Throughout this book, we will get up to speed with Selenium and its surrounding technologies. Selenium is a browser automation tool that has progressed tenfold since its initial inception. Along with tools such as AutoIt, it can be used for automating desktop applications. Selenium is getting used extensively in mobile automation nowadays. The most important point is that it is open source, has a vast developer community, and is constantly evolving. With Selenium Grid, we can simulate different browsers on a single machine.

First, we will start with understanding the basics. This chapter is a gentle introduction to Selenium, and we will be covering the following topics:

- The need for test automation and its advantages
- Java 8 (briefly)
- Selenium RC
- Selenium WebDriver
- Various drivers in Selenium
- Preparing for the first script

Technical requirements

You will be required to have Java 8, Maven, Selenium WebDriver 3.13, Eclipse Kepler or higher.

The code files of this chapter can be found on GitHub:
`https://github.com/PacktPublishing/Selenium-WebDriver-Quick-Start-Guide/tree/`
`master/Chapter01`

Check out the following video to see the code in action:
`http://bit.ly/2PqyMEt`

Why is test automation required?

Let's get started by understanding why test automation is needed. Today's agile world needs quick feedback on the code's quality. The developers check-in application code in a source code repository like GitHub. It is imperative that these changes be tested, and the best way to do so is through automation. A test-automation suite can eliminate the mundane work of manual regression testing and can be helpful in finding bugs earlier, thus reducing manual testing time. It can be configured to run at a particular time in the day.

A cut-off time should be provided to the developers, such as 6 P.M. in the evening, by which time they should check in code, get the application build done, and the application deployed to a server like Apache Tomcat. The automation suite may be scheduled to run at 7 P.M. daily. Jenkins is a tool that's used for continuous integration, and so can be used for this purpose.

Advantages of test automation

Advantages of test automation include reducing the burden on the testers doing the manual execution so that they can focus on the functional aspects of the application. Generally, a smoke, sanity, regression test suite is created for this purpose. The advantage of having automatic triggering through Jenkins is that it facilitates test execution in an unattended mode.

Some pointers on Selenium

We will be using version 3.13, which is the latest version of Selenium at the time of writing this book. It has developed a lot from its early ancestor, Selenium 1. Selenium RC was another tool that would let you write automated web application UI tests in programming languages such as Java, C#, Python, Ruby, and so on, against a HTTP website using any JavaScript enabled browser. For the coding part, we will be working with Java 8. Learning with Java can be fun and at the same time, fast.

What's new in Java 8

Up until Java 7, we only had object-oriented features in Java. Java 8 has added many new features. Some of these features are as follows:

- Lambda expressions and functional interfaces
- Default and static methods in interfaces
- The `forEach()` method in iterable interfaces
- The Java Stream API for bulk data operations on collections

Don't worry if you find this intimidating. We will slowly uncover Java 8 as we progress throughout this book.

Lambda expressions and functional interfaces

Lambda expressions are essential in functional programming. Lambda expressions are constructs that exist in a standalone fashion and not as a part of any class. One particular scenario where Lambda expressions can be used is while creating classes which consist of just a single method. Lambda expressions, in this case, help to be an alternative to anonymous classes (classes without names), which might not be feasible in certain situations. We will briefly look at two examples, side by side, of how we can convert a conventional Java snippet into a Lambda expression.

In the following code, we will assign a method to a variable called `blockofCodeA`. This is just what we are intending to solve with the means of Lambda expressions:

```
blockofCodeA = public void demo(){ System.out.println("Hello World");
}
```

The same piece of code can be written using Lambda expressions, as shown here:

```
blockofCodeA = () -> {
    System.out.println("Hello World");
}
```

Remove the name, return type, and the modifier, and simply add the arrow after the brackets. This becomes your Lambda expression.

Functional interfaces

Functional interfaces contain one—and only one—abstract method. An abstract method is one which should have a body in the implementation class if the implementation class is not abstract. It can have any number of regular methods (methods which have a body in the implementation classes), but the prerequisite of a functional interface is that the number of abstract methods must be only one. These interfaces are used hand-in-hand with Lambda expressions.

In the following code block, the demo method is inside an interface Greeting. Therefore, this interface should only have one abstract method, which is the demo method. In order to instruct other users that this is a functional interface, we annotate this interface with the @FunctionalInterface annotation.

The type of blockofCodeA will be of this functional interface type. This annotation is optional:

```
@FunctionalInterface
public interface Greeting {
        public void demo();
}
```

Default and static methods in an interface

Up until Java 1.7, it was not possible to define a method inside an interface. Now, 1.8 introduces the default methods through which we can provide implementation for a method inside the interface. Let's see an example of this here:

```
interface Phone{
 void dial();
 default void text() {
 System.out.println("Texting a message");
 }
}
```

Static methods in Java are those methods that can be invoked without creating an object of a particular class, provided that the static method is in that particular class. In Java 8, static methods can be defined inside an interface, as shown here:

```
interface Phone {
    inx x;
    void changeRingtone();
    static void text() {
        System.out.println("Texting");
```

```
        }
    }
    public class PhoneDemo {
        public static void main(String[] args) {
            Phone.text();
        }
    }
```

You can invoke the `text()` method directly using the name of the interface.

The forEach method for a collection

Starting with Java 8, we can invoke the `forEach` method on a collection and iterate through the contents of the collection. Let's compare the 7 and 8 versions of iterating over an array list of strings.

The following code, which is from Jave 7, fetches individual fruit names from the fruits list and prints it to the console:

```
    List<String> fruits = Arrays.asList("Apples", "Oranges", "Bananas",
        "Pears");
    for (int i = 0; i < fruits.size(); i++) {
      System.out.println(fruits.get(i));
    }
```

A second alternative that you can use is as follows:

```
   for (String fruit : fruits ){
       System.out.println(fruit);
   }
```

The example shown here does the same thing in Java 8 using lambda expressions:

```
   fruits.forEach(i -> System.out.println(i));
```

Streams in Java 8

As per the Java documentation's definition:

> *Streams are a sequence of elements supporting sequential and parallel aggregate operations.*

Imagine a factory in which workers are standing with tools in their hands, and machine parts keep moving around so that the individual worker can do their part. Streams can be compared somewhat to such a scenario:

```
List<String> fruits = Arrays.asList("Apples","Oranges","Bananas","Pears");
fruits.stream().forEach(fruit -> System.out.println(fruit));
```

Understanding Selenium RC

Selenium RC is a popular UI automation library for automating browsers. Selenium RC uses a generic form of JavaScript called Selenium Core to perform automation. However, this should comply with a security policy called the same-origin policy. The same-origin policy is a security measure that prevents website scripts from accessing the scripts of other websites. For example, JavaScript present on Google cannot access or communicate with JavaScript present on Yahoo. Three things are checked for the same-origin policy: the protocol, domain, and port. If these three things match, then only the request can be said as being one from the same domain.

Selenium Core was introduced by Jason Higgins; It was nothing but a JavaScript program. Prior to Selenium RC, IT people had to install both Selenium Core and the entire web application on their local machine to make the virtual appearance as though the requests were coming from the same domain. Selenium RC introduced the RC server, which acted as a HTTP proxy and handled the requests between the web application and Selenium Core.

What is cross-site scripting (XSS)?

Another concept related to same-origin policy is cross-site scripting. Cross-site scripting refers to the situation where a website can be prone to attacks from hackers. A typical hacker injects one or more JavaScript codes into web pages that are being browsed. These JavaScript codes can be malicious, and can pull cookie information from websites, pertaining to be banks, for example. This way, the malicious script bypasses the same-origin policy control.

Selenium RC consists of two parts:

- Selenium server
- Client libraries

The following diagram shows the functioning of Selenium RC, where the RC Server sits in-between the libraries like Java and Python and sends instructions to Selenium Core, thereafter operating on the individual browser:

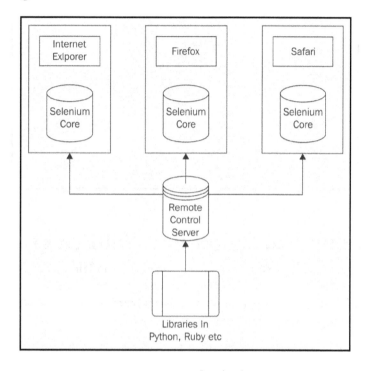

Image modelled from www.seleniumhq.org

The role of the **Remote Control Server** is to inject the **Selenium Core** in the respective browser. The client libraries send instructions in the form of requests to the RC Server, and the RC Server communicates this to the browser. After receiving a response, this is communicated back to the user by the RC Server.

Introducing Selenium WebDriver

Selenium WebDriver is used for automating web browsers by using the browser's internal plugins or dll with the individual browser drivers, which are available for each individual browser.

The following diagram shows the high-level functionality of Selenium WebDriver. The JSON API parses the instructions from languages like Java, Python, and so on, and invokes and operates on the concerned browser:

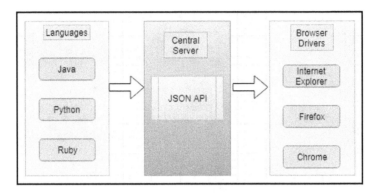

Class structure of Selenium WebDriver

The following diagram is a snapshot of the class structure of Selenium WebDriver. The WebDriver interface is the parent of Remote WebDriver, which is a public class. Drivers for Internet Explorer, Firefox, Chrome, and so on inherit from the Remote WebDriver. In future chapters, we will be digging deep into these drivers:

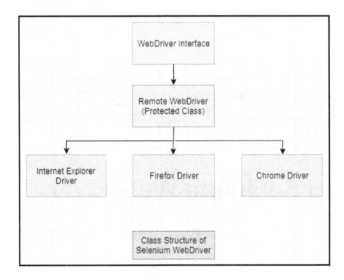

Drivers in Selenium

We will now take a look at the various drivers that are available in Selenium and their usage.

Remote WebDriver

Remote WebDriver is the implementation class of the WebDriver interface. Apart from WebDriver, it also implements the interfaces of `TakesScreenShot`, `findBy`, `JavaScriptExecutor`, and so on.

Mobile drivers

All modern web apps have implementations for mobile devices. The two most popular operating systems in mobile devices are Android and iOS. Selenium has implementations for Android and iPhone, that is, AndroidDriver and IODDriver. Both of these are direct implementations of WebDriver.

Headless browsers

Headless browsers are those that do not have a **graphical user interface** (**GUI**). Everything runs in the background. When a test is executed with a headless browser, no screen is displayed to the user. Two popular headless browsers are HTMLUnit and Phantom JS. Chrome now supports the HTMLUnit browser.

Why do we need headless browsers?

Suppose that Selenium tests have to be executed on an OS which does not have a GUI like Linux or when multiple browser behaviors have to be simulated on just one machine. The advantage of a headless browser is that the resources utilized by the test are minimal. A scenario where you can use these browsers is for test data creation. In these situations, there is no special need to display the screen to the user.

Preparing for the very first script

Follow the steps shown here to get started with Selenium WebDriver.

Installing Java 8

Follow the instructions below to install Java 8:

1. Go to http://www.oracle.com/technetwork/java/javase/downloads/jdk8-downloads-2133151.html and click on the appropriate version. I have selected the 64-bit Windows version since mine is a Windows machine.

2. Once the file has downloaded, run the .exe file. Java will start installing it onto your machine. Next, we have to set two environment variables in order to use Java.

3. Go to **Control Panel** and click **Advanced System Settings**.

4. Click on **Environment variables** and add two system variables:
   - One is JAVA_HOME. Provide the path of the root folder where Java is installed. In this case, this will be C:\Program Files\Java\jdk1.8.0_152.
   - The second is the Path variable. Remember that this variable has to be appended after adding a ;. Here, the path of the bin folder has to be specified. In this case, this will be C:\Program Files\Java\jdk1.8.0_152\bin.

5. The next step is to check our configuration. Open the Command Prompt and type java –version:

If you get an output similar to the one shown in the preceding screenshot, you are all set to start coding.

Now, let's get our hands dirty!

Setting up Eclipse

We will be using Eclipse as an IDE for developing Selenium Scripts in this book, but you are free to use whichever IDE suits you best.

Downloading Eclipse

Navigate to the Eclipse website (`www.eclipse.org/ide`) and click on the **Download** link. Here, you can find very specific instructions regarding how to install you favourite IDE version (Kepler, Neon, and so on).

Creating a Maven project

Once the IDE is installed, perform the following steps:

1. Double click on the `.exe` file for Eclipse and go to **File | New | Other**.
2. Select **Maven Project**. Click **Next**.
3. Click **Next** on the screen that appears.
4. Select **Create a simple project (skip archetype selection)**. Then, click **Next**.
5. Input the **Group ID**. Ideally, this is the package name of the project. The **Artifact ID** corresponds to the name of the JAR file in case you want to create one. Keep the packaging as JAR. Notice that the version is 0.0.1-SNAPSHOT. The SNAPSHOT part indicates that the project is still under development and has not been released.

6. Click **Finish**. The following is a snapshot of the **Project Explorer**:

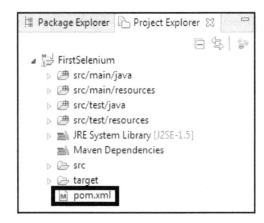

When you create a **Maven Project**, the `src/main/java`, `src/main/resources`, `src/test/java`, and `src/test/resources` folders, are created for you. Apart from these, you will see a `Maven Dependencies` folder that is currently empty. Marked with a black box, there is a `.xml` file called `pom.xml`. This is the place where you will place all of the dependencies for your project. By dependencies, I mean dependent JARs. JAR stands for Java archive.

Understanding pom.xml

It's time to explore `pom.xml`. This is what `pom.xml` looks like:

```
<project xmlns="http://maven.apache.org/POM/4.0.0"
xmlns:xsi="http://www.w3.org/2001/XMLSchema-instance"
xsi:schemaLocation="http://maven.apache.org/POM/4.0.0
http://maven.apache.org/xsd/maven-4.0.0.xsd">
 <modelVersion>4.0.0</modelVersion>
 <groupId>org.packt.selenium</groupId>
 <artifactId>FirstSelenium</artifactId>
 <version>0.0.1-SNAPSHOT</version>
</project>
```

Group ID and Artifact ID that you added in the previous screens have appeared in the preceding file, inside the `Project` tag. In order to work with Selenium, we will need to add Selenium dependencies within the Project tag. Let's go ahead and add those from the Maven repository:

1. Go to the Maven repository (`https://mvnrepository.com`) and grab the dependency shown here:

```
<!--
https://mvnrepository.com/artifact/org.seleniumhq.selenium/
selenium-java -->
<dependency>
    <groupId>org.seleniumhq.selenium</groupId>
    <artifactId>selenium-java</artifactId>
    <version>3.13.0</version>
</dependency>
```

2. Place this dependency inside a `dependencies` tag, as shown in the `pom.xml` file here:

```
<project xmlns="http://maven.apache.org/POM/4.0.0"
xmlns:xsi="http://www.w3.org/2001/XMLSchema-instance"
xsi:schemaLocation="http://maven.apache.org/POM/4.0.0
http://maven.apache.org/xsd/maven-4.0.0.xsd">
    <modelVersion>4.0.0</modelVersion>
    <groupId>org.packt.selenium</groupId>
    <artifactId>FirstSelenium</artifactId>
    <version>0.0.1-SNAPSHOT</version>
    <dependencies>
        <!--
https://mvnrepository.com/artifact/org.seleniumhq.selenium/
selenium-java -->
        <dependency>
            <groupId>org.seleniumhq.selenium</groupId>
            <artifactId>selenium-java</artifactId>
            <version>3.13.0</version>
        </dependency>
    </dependencies>
</project>
```

3. Save the `pom.xml`. You will see a small activity in the bottom-right corner of Eclipse, stating that the project is being built.

The `Maven Dependencies` folder now gets populated with all of the downloaded JARs, as shown previously.

Manual configuration

With this, we are ready, and have the basic Eclipse setup for Selenium WebDriver. But we are not done yet. It might occur that, under a corporate firewall, you are unable to download the required JARS. In this situation, perform the following steps:

1. Simply create a plain **Java** project.
2. Right -click on the project in **Project Explorer.**
3. Select **Build Path | Configure Build Path.**
4. Click on **Add External JARs** and add the required JARs manually.
5. Next, we will write a very simple script which just opens www.google.com (this is shown in the following section). Right-click the **Project** and select **new class.**

Creating the first script

Type the following code. What the following script does is simply opens a new Chrome browser and navigates to the URL http://www.google.com:

```
public class FirstTest {
    public static void main(String[] args) {
        System.setProperty("webdriver.chrome.driver",

"C:\\SeleniumWD\\src\\main\\resources\\chromedriver.exe");
        WebDriver driver = new ChromeDriver();
        driver.get("http://www.google.com");
    }
}
```

Right-click the file and click **Run as Java Application** and hurrah! A chrome browser opens and http://www.google.com gets loaded.

You have successfully created your first Selenium Script.

Summary

This chapter gave you an idea of what Selenium RC and WebDriver are and also touched upon concepts like same-origin policy and cross-site scripting. We also did a basic Eclipse setup using Maven as well as without Maven, and finally we created a very simple program to open a URL in the browser.

In Chapter 2, *Understanding the Document Object Model and Creating Customized XPaths*, we will learn about the Document Object Model and its various traversal techniques.

2
Understanding the Document Object Model and Creating Customized XPaths

In the previous chapter, we had a look at Selenium RC, WebDriver, the basic Eclipse setup, and a very simple program. Opening the browser and loading a web page is OK to start with but it is not of much use if one is not able to do operations such as clicking links and buttons, selecting values from dropdowns, entering text in a textbox, and so on. In order to achieve this, we have to explore what is called the **Document Object Model** (**DOM**). The following topics will be covered in this chapter:

- What is the DOM?
- DOM traversal
- What is the Fillo API?
- Debugging in Eclipse

The four major topics will be explained in detail in the form of sub-topics throughout the chapter.

Technical requirements

You will be required to have Java 8, Maven, Selenium WebDriver 3.13, Eclipse Kepler or higher and Google Chrome.

The code files of this chapter can be found on GitHub:
https://github.com/PacktPublishing/Selenium-WebDriver-Quick-Start-Guide/tree/master/Chapter02

Check out the following video to see the code in action:
`http://bit.ly/2PY4oP1`

What is the DOM?

The DOM is an application programming interface that is linked to HTML, XHTML, or XML documents and treats these similar to a tree where each node in the tree represents a part of the document.

In simple words, the DOM can be compared to a tree where there is a root node, intermediate nodes, and leaf nodes.

 The root node has no parent; the intermediate nodes have a parent, one or more siblings, and one or more children. This is a very important concept and will help at the time of creating relative or customized XPaths, which we will see in a later part of the chapter.

Shown here is a sample DOM:

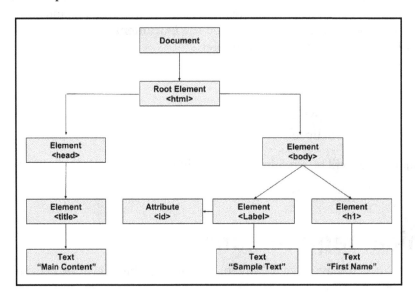

In the preceding diagram, there are three leaf nodes, two intermediate nodes, and one root node. All of this is contained in a document. The question that arises next is: how do we get to a particular node in this tree structure? This is where terms such as XPath and CSS come into picture.

XPath stands for XML Path and is a querying language to query the nodes in a DOM.

CSS stands for **Cascading Style Sheets** and is a style sheet language used for describing the look and format of a document written in any markup language.

WebElements

Any HTML element on the page, such as a textbox, dropdown, radio button, checkbox, and so on can be considered as a `WebElement`. How can we access web elements on an HTML screen? Well, Selenium WebDriver provides us with the very useful `WebElement` interface. The only implementation class of this interface is the `RemoteWebElement`. Whenever an HTML element has to be operated upon in the code, it's a best practice to access the element using the `WebElement` interface. One of the parent interfaces of the `WebElement` interface is `SearchContext`, which we will look at next.

SearchContext interface

`SearchContext` is the interface that has been designed to find elements in the DOM using a locator mechanism. `SearchContext` has two public methods:

- `findElement`: This identifies all matching elements based on a locator mechanism but always returns the first matching web element.
- `findElements`: This returns a list of `WebElements` that match the locator.

When Selenium returns a `WebElement`, it does not return an object of the `WebElement`, since the `WebElement` is an interface. Rather, it returns an object of `RemoteWebElement`, which is the implementation class of the `WebElement` interface.

What if the locator mechanism does not return any elements? In this case, `findElement` will always throw a `NoSuchElementException`. The `findElements` method, on the other hand, will not throw a `No SuchElementException`. It returns an empty list.

> Always use `findElements` when unsure about the existence of a `WebElement` based on some locator mechanism. This is the best practice to avoid `NoSuchElementException` at runtime. It is very likely you'll get this exception while using the `By.tagName` static method, which we will see later.

DOM traversal

Now that we have seen the basic definitions of XPaths and CSS and also looked at what WebElements are, let's understand how we can use these to exploit the DOM. DOM traversal entails getting to the desired element with the help of either XPaths or CSS. It is possible to traverse the DOM in a forward and backward direction with XPaths but traversal through XPaths is slow compared to CSS. Traversal using CSS can only be done in the forward direction. In order to traverse the DOM, using either XPaths or CSS, we need to understand the By class.

Dissecting the By class

The By class is an abstract class that has eight static methods and eight inner classes. Let's understand the structure of the By class.

The following code skeleton shows a fragment of the structure of the By class:

```
public abstract class By {
        public static By id(String id);
        public static By cssSelector(String css);
        public static By name(String name);
        public static By linkText(String text);
        public static By className(String className);
        public static By tagName(String tagName);
        public static By partialLinkText(String partialLinkText);
        public static By xpath(String xpath);
        public static class ById extends By {
                WebElement findElement(By by);
                List <WebElement> findElements(By by);
    }
}
```

 Note: Inner classes are present corresponding to all the static methods. There are inner classes such as ByName, ByTagName, and so on.

Inner classes similar to ById exist for name, linkText, xpath, and so on. We will be using the two static methods xpath and cssSelector to design what are called customized xpath and css. Let's try to understand the various mechanisms to access the DOM elements. There are eight ways to access WebElements using the static methods in the By class. We will just be covering access using static methods.

We will look at each of the eight methods individually, and then we will adopt a better approach using the relative (customized) XPath, which will cover each method internally:

- `By.id`: Uses the `id` attribute of the element to locate. For example, `By.id("userid")`.
- `By.name`: Uses the `name` attribute of the element to locate. For example, `By.name("username")`.
- `By.className`: Uses the `class` attribute of the element to locate. For example, `By.className("class1")`.
- `By.linkText`: Uses the text of any anchor link to locate. For example, `By.linkText("Click here to Login")`.
- `By.partialLinkText`: Uses the partial text of any anchor link to locate. For example, `By.partialLinkText("Login")`.
- `By.xpath`: Uses the XPath of the element to locate. For example, `By.xpath("//*[text()='Login']")`.
- `By.cssSelector`: Uses CSS selectors to locate. For example, `By.cssSelector(".ctrl-p")`.
- `By.tagName`: Uses tag names, such as `input`, `button`, `select`, and so on to locate. For example, `By.tagName("input")`.

The two types of XPaths

Let's now understand what absolute XPaths are and how they differ from relative (customized) XPaths:

- An absolute XPath is the entire path of the `WebElement` taken from the root node. For example, `html/body/div/a`.
- A relative or customized XPath is one in which we use the following format. For example, if the `div` has an `id` of `ABC`, then the same absolute XPath will be `//div[@id='ABC']/a`.

There is an apparent problem with the absolute type of XPath. If the DOM structure changes in the future (for example, if `div` is removed), then this path will undergo changes.

Understanding customized XPaths

The structure of a customized XPath is given as follows: `//*[@Attribute = 'Value']`.

Here, `//` indicates that the entire DOM will be searched. We will understand some important XPaths with the help of `http://www.freecrm.com`:

Mentioned below are some of the commonly used strategies.

- Using the name attribute: `//*[@name='username']`. This searches the DOM for an element for which the name value is `username`. This is the login field on the landing page.
- Using the name and type attributes: `//*[@type='password'][@name='password']`. In the DOM, the `password` field on the screen can be identified using just the `name` field. Just for the sake of demonstrating multiple attributes, I have taken the type attribute also. The need for multiple attributes arises when a unique element cannot be found using just one attribute.
- Using the `contains` clause: `//*[contains(@type, 'password')]`. This searches for an element whose type attribute contains the text `password`.
- Using `starts-with`: `//*[starts-with(@name, 'user')]`. This XPath will find the `username` field again but this time based on the starting text present in the `name` attribute.
- Using the following node: `//*[@name='username']//` following the `::` input. This XPath searches for input tags which follow the username field. The boundary of this search is the container element within which the username lies. Since there is a password textbox and **Login** button following the username and the username, password, and **Login** button are inside a form, it identifies the password textbox and **Login** button.
- Using the node: `//*[@value='Login']//` preceding the `::` input. This will provide the `username` and `password` textboxes.
- Using the `onclick` attribute: `//a[contains(@onclick, 'html/entlnet/userLogin.html')]`. This is a very common case and is used when we have anchor tags without an ID or name and just an `onclick` attribute that has a JavaScript function called `onclick()={function content}`. In this case, the anchor tag can be structured as `<a onclick='window.open("http://boot1/html/entlnet/userLogin.html")'>Login</a>`.

- Using the `ExtJs qtip` attribute: `//*[@*[local-name()='ext:qtip'][.='Account Number']]`. With the growing popularity of `ExtJs` for developing web apps, it is necessary to have something to identify common `ExtJs` attributes. One `ExtJs` attribute is `qtip`. Here we are finding an element with the `qtip Account Number`.
- Using `and`: `//input[@class='textboxes' and @name='firstName']`. In this case, an input element with the class attribute as `textboxes` and name as `firstName` will be located. Both conditions in `and` must be satisfied.
- Using `or`: `//input[@class='textboxes' or @name='firstName']`. In this case, an input element with the class attribute as `textboxes` or name as `firstName` will be located. Either of the conditions in `and` must be satisfied.

Customized CSS

Now that we have seen the customized XPath, it's time to look at customized CSS. Remember, CSS can be used only for forward traversal.

The following are some customized CSS examples that one can use while coding the program:

- Using the name attribute: `input[name='username']`. This CSS identifies the username. Notice there are no '`//`'s.
- Using the name and type attribute: `input[type='password'][name='password']`. This will identify the `password` textbox.
- Using the `ID` and `class`: `form[id='loginForm']`, `form[id='loginForm'][class='navbar-form']`. These two CSS selectors will identify the login form.
- Using the '`contains`' clause: `form[id*='Form']`. This will identify the form since the `ID` of the form contains the text `Form`. Contains is indicated by '`*`' in CSS.
- Using the '`starts-with`' clause: `form[id^='login']`. This will identify the form since the form `ID` starts with the text '`login`'. `starts-with` is indicated by '`^`' in CSS.
- Using the '`ends-with`' clause: `form[id$='Form']`. This will identify the form since the form `ID` ends with the text '`Form`'. `ends-with` is indicated by '`$`' in CSS.

An example traversal

The element retrieval and traversal can be done quite easily by what is known as a browsers console. In all the three browsers, the console can be invoked by pressing the *F12* key on the keyboard. In Chrome, the **Elements** tab will help in finding the XPath. One can traverse back and forth in the DOM using '/..' and ('//' or '/'). Let's see what the Chrome console looks like.

The following snapshot shows the Chrome console with the username field highlighted because we tried to find an element through it's XPath . In order to search for any element, just press *Ctrl + F* on the console. A search box opens where you can type the XPath:

A similar console in Internet Explorer is called **Developer Options** and in Firefox it is called **Firepath**. In Firefox, one must remember to first add the firebug plugin from the Firefox plugins page (go to the **Tools | Add-ons** menu and then select **Add-ons** from the left pane). Only then can Firepath be accessed using the *F12* key.

Apart from the consoles, which come built-in with the browsers, there are a few extensions such as XPath helper in Chrome and MRI in Internet Explorer. MRI is a bookMarklet for IE. One can get it from `http://westciv.com/mri/` as a free installation. All the instructions are available on this website.

 MRI will not work on popup windows. In the case of popups, the console is a better option.

Understanding the text() methods

One very useful method in finding XPath is the `text()` method. When we need to supply some text at runtime, say for example, from an Excel file, then we can utilize the `text()` method in the following manner:

```
public class DynamicText {
   public static void main(String[] args) {
     System.setProperty("webdriver.chrome.driver",
        "C:\\SeleniumWD\\src\\main\\resources\\chromedriver.exe");
     WebDriver driver = new ChromeDriver();
     driver.manage().timeouts().implicitlyWait(30, TimeUnit.SECONDS);
     driver.get("http://www.google.com");
     String variableData = "Google";
     String dynamicXpath = "//*[text()='" + variableData + "']";
     List<WebElement> elem =
     driver.findElements(By.xpath(dynamicXpath));
     System.out.println("no of elements: " + elem.size());

   }
}

The program above prints
no of elements: 3
```

Finding elements within the container element

On the `http://www.freecrm.com` login page, the structure is such that the username, password, and the **login** button are contained inside the form with id=xyz. In such a situation, the child elements can be accessed using `findElements` on the container or parent element.

The following code displays the number of input elements in the form:

```
public class DynamicText1 {
 public static void main(String[] args) {
    System.setProperty("webdriver.chrome.driver",
    "C:\\SeleniumWD\\src\\main\\resources\\chromedriver.exe");
    WebDriver driver = new ChromeDriver();
    driver.manage().timeouts().implicitlyWait(30, TimeUnit.SECONDS);
    driver.get("http://www.freecrm.com");
    String dynamicXpath = "//*[@id='loginForm']";
    List<WebElement> elem =
   driver.findElements(By.xpath(dynamicXpath));
    List<WebElement> elem1 =
    elem.get(0).findElements(By.tagName("input"));
    System.out.println("no of elements: " + elem1.size());
    }
}

The output displayed in console is shown below
no of elements: 3
```

This is a very simple program which has hardcoded values. To remove hardcoding from a program, we require a framework, which we will discuss in forthcoming chapters.

Best practice

A best practice while coding Selenium is always to follow a design pattern. We will go over design patterns in a subsequent chapter.

We should always have modular code delinked from each other so that when one module changes, there is no impact on other modules.

Extracting WebElements dynamically using tagName

Now that we have seen how to create relative (customized) XPaths, it's time to see how to retrieve WebElements programmatically using Java lists. The best way to understand this is through an example. Suppose we want to find all the input textboxes on the login page of http://www.freecrm.com. We will make use of the findElements method. Remember, the findElements method is in the SearchContext interface.

Since the WebDriver interface is a child interface of `SearchContext`, it inherits the `findElements` method and we can invoke this method on the reference variable of WebDriver. In conjunction with `findElements`, we will also make use of the static method, `tagName` , of the `By` class.

The following code makes efficient use of the list interface in Java (present in the `Java.Util` package):

```
public class URLTest {
 public static void main(String[] args) {
     System.setProperty("webdriver.chrome.driver",
         "C:\\SeleniumWD\\src\\main\\resources\\chromedriver.exe");
     WebDriver driver = new ChromeDriver();
     driver.get("http://www.freecrm.com");
     List<WebElement> inputBoxes =
     driver.findElements(By.tagName("input"));
     System.out.println("No of inputBoxes: " + inputBoxes.size());
 }
}

Output from this program:
No of inputBoxes: 3
```

The two textboxes for `UserName` and `Password` and the **Login** button are treated as input tags. The `tagName` static method is an extremely useful method and you can use this method for almost any element on any web page.

Properties file for WebElements

We have explored `WebElements` to a large extent. Now we will actually start preparing for the hybrid framework (we will look at this in a later chapter) by creating a `WebElement` store. This store will be created in a file known as the properties file, which always has the `.properties` extension. An example entry in the properties file can be:

```
USERNAME=//*[@name='username']
```

Entries in the properties file consist of key value pairs. Here, `username` is the key and `'//*[@name='username'] '` is the value.

> The key in a properties file should always be unique. The value part can have duplicate values.

These values should be retrieved by the code once the key is supplied. For this purpose, we will be writing a retrieval program in a subsequent chapter.

The next question that might come up to mind is: we have created a properties file and will be writing retrieval logic for this, but on what basis should the retrieval logic be invoked? For this purpose, we will have to create test scripts. The test scripts can be created either in an Excel or in a database. We have APIs such as Apache POI and Fillo available as open source. Fillo gives us certain advantages over POI. Fillo treats an Excel tab as a database table and regular SQL queries, such as SELECT, UPDATE, and DELETE, can be triggered on the Excel tab data. Each row is equivalent to the row in a DB table while a column is equivalent to a database field. We will gradually uncover the power of Fillo as we move ahead but the curtain raiser will be in Chapter 3, *Basic Selenium Commands and Their Usage in Building a Framework*. As of the current release of Fillo, joins are not possible but we will not require complex joins for our framework.

Let's take a small diversion here to see what the prerequisites for automating mobile applications are.

Prerequisites for automating mobile applications

For automating mobile applications, there is specific software that needs to be downloaded. The following is a list of all the steps needed to set up the Appium server on your machine:

1. **Download the Java Development Kit (JDK)** (http://www.oracle.com/technetwork/java/javase/downloads/jdk8-downloads-2133151.html).
2. Set the Java environment variable path so that Java commands can be executed from anywhere on the system.
3. Download the Android SDK/ADB (https://developer.android.com/studio/).
4. Utilize the Android SDK packages from SDK Manager in the downloaded Android SDK folder
5. Set the Android Environment variables so that commands can be executed anywhere on the system.
6. Download and configure nodejs (https://nodejs.org/en/download/). Take the LTS for whichever OS is applicable.
7. Download the Microsoft .net framework (https://www.microsoft.com/en-in/download/details.aspx?id=30653).
8. Download the Appium server (http://appium.io/).

XPaths for mobile applications

Let's first understand the various types of mobile applications. There are three types of mobile applications:

- **Web application**: Works only in the mobile browser (for example, a personal blog site)
- **Native application**: Works only as a standalone app (for example, a calculator)
- **Hybrid application**: Works on a mobile browser or standalone (for example, Gmail, Flipkart, and so on); it can also be defined as an application that contains a native view and web view

Finding XPaths for mobile browser applications

We have a variety of ways in which we can find the locators of mobile elements. Let's explore some of them.

Connecting the actual mobile device

Perform the following steps to find locators using an actual device connected to the computer:

1. Type `chrome://inspect/#devices` and ensure that **Discover USB devices** is checked.
2. Type URL in the URL textbox and click **Open**. The website now opens in the connected device.
3. Click **Inspect** in Chrome on the desktop. A new instance of Chrome Developer tools opens in the desktop.
4. We can interact with mobile web elements using DevTools.

How to use Screencast

Follow the steps to use screencast in DevTools:

1. Perform the preceding steps 1-4.
2. Click on the screencast icon in DevTools.

A window opens in which one can see the URL opened in the mobile device. You can interact with this window using DevTools.

Appium Inspector window

To use the **Appium Inspector** window, perform following the steps and find the desired locator:

1. Start the Appium server.
2. In the downloaded `Android SDK` folder, open the **Android AVD (Android Virtual Device)** Manager.
3. Start the emulator inside the AVD.
4. Click the magnifying glass icon on the Android server GUI. This opens up the **Android Inspector**.

How to use UIAutomatorViewer

Perform the following steps to use **UIAutomatorViewer** for capturing the locator:

1. Download Android SDK from `https://android-sdk.en.lo4d.com/`.
2. Once downloaded, go to `Andriodsdk/tools` and double-click `uiautomatorviewer.bat`.
3. Click on the device screenshot button, second from the left. The device image gets displayed in the left pane.
4. Click on any element and the corresponding information is displayed in the right pane.

Mobile locators

The main locators used in mobile automation are as follows:

- `Accessibility id`: Unique identifier for a UI element.
- `TagName`: The same as WebDriver. This tells us what the tag is (input, select, and so on).
- `Class Name`: Identifies by the `classname` attribute.
- `Xpath`: Identifies by the absolute or customized XPath.
- `ID`: Identifies by the ID of the element.

What is a WebView?

The browser view that is embedded inside a native app is called a Web View. To view the XPath of a webview in a hybrid app, we make use of the **Selendroid Inspector**. To use the **Selendroid Inspector**, perform the following the steps:

1. Open the **Appium Server** GUI and put the local APK file path in the application path.
2. Select the **Automation Name** as `Selendroid` and the other mandatory parameters.
3. Start the server after selecting **Pre-Launch Application**.
4. Navigate to `http://localhost:8080/inspector` and start using the **Selendroid Inspector**.

Introducing the Fillo API

In the subsequent chapters, we will be dealing with a lot Excel data. In order to make data manipulation simpler using Excel, **Codoid,** a software firm, introduced the Fillo API. This API can be used for both XLS and XLSX files. It's an open source API. Fillo can be used to trigger select, insert, and update operations with `where` conditions. It also supports `like` and multiple `where` conditions. Let's understand how to make use of this useful API. First of all, let's add the Maven dependency for Fillo. Put the dependency shown here in `pom.xml`:

```
<dependency>
    <groupId>com.codoid.products</groupId>
    <artifactId>fillo</artifactId>
    <version>1.18</version>
</dependency>
```

To start using `Fillo` in the code, create a class file in Eclipse and type the following code in a `main` method. This code first creates a `Fillo` object using the `Fillo()` no-argument constructor. After that it creates a `Connection` object. The value assigned to this object is obtained from the `getConnection()` method, which takes the full file path of the `.xls` or `.xlsx` file as an argument:

```
package org.packt.selenium;

import com.codoid.products.exception.FilloException;
import com.codoid.products.fillo.Connection;
import com.codoid.products.fillo.Fillo;
import com.codoid.products.fillo.Recordset;
```

```java
    public static void main(String[] args) throws FilloException {
      Fillo fillo = new Fillo();
      Connection conn = null;
      try {
        conn = fillo
.getConnection("D:\\workspace\\hdfc1\\src\\test\\testAutomation\\resources\
\Framework.xls");
      } catch (FilloException e) {
        throw new FilloException("File not found");

      }
      String query = "Select * from TestConfig where Execute_Flag='Y'";
      Recordset rcrdset = null;
      try {
        rcrdset = conn.executeQuery(query);
      } catch (FilloException e) {
        throw new FilloException("Error executing query");

      }
      try {
        while (rcrdset.next()) {
          System.out.println(rcrdset.getField("TestCaseID"));
        }
      } catch (FilloException e) {
        throw new FilloException("No records found");

      } finally {
        rcrdset.close();
        conn.close();
      }
    }
```

This piece of code will print out
TestCase1
TestCase3
TestCase4

The `Framework.xls` has the following content. As can be seen here, in the `TestConfig` tab, there are four test cases, out of which one has `'N'` set in the `Execute_Flag` field. Hence, only three test case IDs get printed in the console:

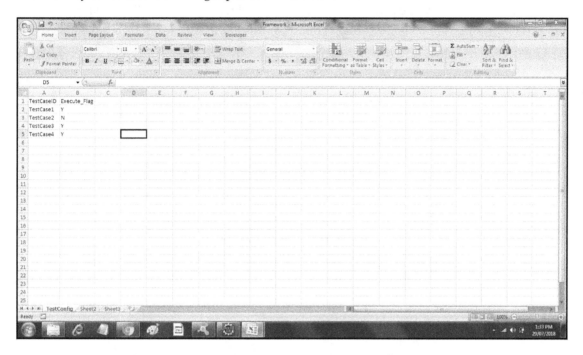

Notice that in the preceding code, we have hardcoded the path to the `Framework.xls` file. In `Chapter 7`, *The Command Pattern and Creating Components*, we will be creating a `config.properties` file, which will hold config level details, such as the URL and framework file path. The idea of having property files is to have config or object repository-level data at a central location. This helps when tests have to be done in different environments from time to time.

`Update` is another common function that can be done using `Fillo`. We will look at `update` when we start creating the framework.

Debugging in Eclipse

During the coding journey, a very useful technique to figure out logic problems is known as **debugging**. You remove the bugs in the code using debugging. To use the debugger in Eclipse, navigate to the **Run | Debug As** menu, or simply press *F11*:

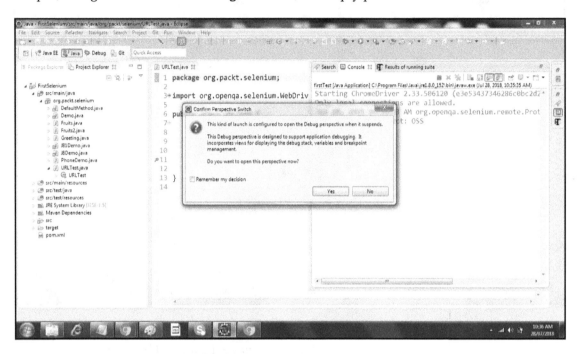

You will get the popup shown previously. Optionally, click on **Remember my decision** and then click on the **Yes** button.

The screen that gets displayed is shown as follows:

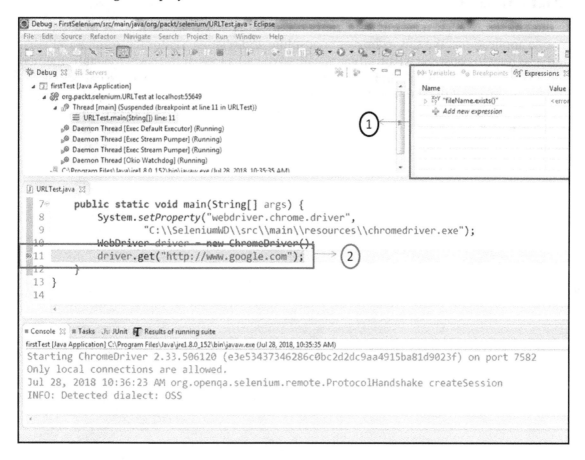

The box labelled 1 is where the variable and expression values can be checked and the box labelled 2 is the breakpoint on which the execution has halted. Now one can progress line-by-line using *F5* or *F6*. *F5* will go into each individual called method, whereas *F6* will not. Clicking *F8* will continue the execution, which might halt at the next breakpoint, if present.

The topic of debugging can consume an entire mini-book but as far as this book is concerned, this much knowledge about debugging should be sufficient enough.

Summary

In this chapter, we had a look at the different locators available for desktop as well as mobile applications. We saw examples of absolute and relative XPaths (customized XPaths) and also how to construct XPaths dynamically. We had a look at the `tagName` static method and also understood how to find elements with container elements. Next, we had a look at mobile locators and the different tools available to us for automating mobile applications. We understood the different types of mobile applications. Finally, we touched on debugging in the Eclipse IDE. In `Chapter 3`, *Basic Selenium Commands and Their Usage in Building a Framework*, we will start looking at writing pieces of code in a linear fashion.

3
Basic Selenium Commands and Their Usage in Building a Framework

In the last chapter, we learned what the **Document Object Model** (**DOM**) and DOM traversal are and how they can be done using XPath and CSS. We also saw the various static methods, such as `tagName()`, and explored the `size()` method to verify the number of web elements fetched by `findElements`. Moreover, we saw how to find web elements within container elements and the various tools available for DOM traversal.

In this chapter, we will be covering the following topics:

- Understanding method signatures
- `findElement` and `findElements` with the concept of List
- Various methods that can be used for browser navigation
- Methods that can be used on a web element that has been found using `findElement` or `findElements`
- Creating reusable methods for textboxes, radio buttons, checkboxes, and drop-downs, among others
- Extracting elements dynamically
- Extracting elements from container elements
- A brief look at the framework architecture with the required components, such as Log4J and Extent Reports

Technical requirements

You will be required to have Google Chrome, Eclipse Kepler or higher, Java 8, Selenium WebDriver 3.13, Log4J 1.2.17, Extent Reports 3.1.5 and Microsoft Excel or Open Office.

The code files of this chapter can be found on GitHub:
`https://github.com/PacktPublishing/Selenium-WebDriver-Quick-Start-Guide/tree/master/Chapter03`

Check out the following video to see the code in action:
`http://bit.ly/2RjPhjb`

What are method signatures?

A method signature can be very simply defined as the method name followed by the parameter types. Say, for example, we have a method that adds two numbers. This is the method signature of this kind of method:

```
addNum(int, int)
```

The actual method is shown in the following code block. This method has to be invoked from another method or from the `main()` method, as in the following:

```
public int addNum(int num1, int num2) {
    int result = num1 + num2;
    return result;
}
Method call: addNum(10,20)
Output: 30
```

In the sections that follow, we will be looking at the method signatures of various important methods. But first, let's understand a concept of Java called List, which we will see in the `findElements` method.

What are Lists in java?

List is the child interface of Collection. It is a part of the Collection framework. List is an ordered collection of objects in which duplicate values can be stored. Since List is an interface, we need to use the implementation class to create an object. We will be using the `ArrayList` class for this purpose. The way to create an object for a List is shown here:

```
List simpleList = new ArrayList(); or
```

```
List<String> simpleList = new ArrayList<String>();
```

The second listing contains generics. Here, we specify that our `List` is going to contain only strings, whereas the first version specifies that our list will contain any kind of object.

Important methods in Selenium

We will have a look at a few of the very important and frequently used methods and their signatures.

These are methods invoked by the WebDriver instance:

- `findElement(By)` versus `findElements(By)`: Both these methods take an argument of type `By`. The `findElement` method calls the `FindElements` method and returns a `WebElement`. If multiple elements are found with the same locator, it returns the first `WebElement`. The `findElements`, on the other hand, returns a list of `WebElements`. The `findElement` throws a `NoSuchElementException` if it is unable to find an element using the locator provided in the parameter list. The `findElements` returns an empty list if no elements are found using the given locator.
- `get(String)`: The `get` method takes a String URL as a parameter. It navigates to the URL supplied as a parameter. The URL can be in the form of `http://www.google.com`. If the `http://` is not included, then an `Unhandled inspector error` gets thrown with the message `Cannot navigate to invalid URL`.

- navigate(): The navigate method is present in the WebDriver interface. This method returns an object of type Navigation, which is a nested interface of WebDriver interface. We get the Navigation object by invoking the navigate() method on the WebDriver object. There are four methods in the Navigation class that are used frequently, namely: to(java.lang.String), forward(), back() and refresh(). The to(java.lang.String) method is used to load a particular webpage in the browser. The overloaded version of the to method is to(java.net.URL). This overloaded version makes it simple to pass a URL. Using back(), one can move back one page in the browser's history. Using forward(), one can move forward one page and refresh() refreshes the current page.

These are methods invoked at the WebElement level:

- sendKeys(CharSequence...): This method sends a character sequence to an element when invoked on the findElement method. The findElement method returns a WebeElement, after which we invoke sendKeys, since sendKeys is a method in the WebElement interface. It can also be invoked on findElements after fetching an individual element from the ArrayList of elements using the get(int) method.
- click(): This method is used to click on a WebElement—the like button, radio button, or any other kind of button. The prerequisite of using the click method is that the element should be visible and should have a height and width greater than 0.
- submit(): If the current element is a form or an element within a form, the information will be submitted to the remote server. If the current element is not within a form, the NoSuchElementException is thrown.
- selectByIndex(int): Selects the option at the index equal to the argument value.
- getTagName(): Gets the tagName of the WebElement returned by the findElement(By) method.
- getText(): Gets the visible text of the element, including subelements.
- getAttribute (String): Fetches the value (which is equal to the string parameter) of the attribute of the WebElement.
- getCssValue(String): Fetches the value of the CSS property that is passed in when invoked on a WebElement.
- getSize(): Fetches the width and height of the rendered element.

- `getLocation()`: Returns a point that has the location of the top-left corner of the element.
- `isDisplayed()`: Returns `true` if the element is displayed.
- `isSelected()`: Returns `true` if this element is selected. This applies only to options in a `Select` object, checkboxes, and radio buttons.
- isEnabled(): Returns true if this element is enabled.

This method is invoked by the `Select` object:

- `selectByVisibleText(String)`: Used on a drop-down list. It selects the item from the list, the text of which matches the string argument.

Some common reusable methods

Now that we know most of the frequently used methods with their signatures, let's create a few common reusable methods. We will create methods for `sendKeys`, `click`, `selectByVisibleText`, `getText`, and `isDisplayed`. These methods have been named as `enterText`, `clickElement`, `selectValue`, `getElementText`, and `isElementDisplayed` respectively. I have taken a few representatives. As an exercise, you can create similar methods as needed. These are called wrapper methods. Here, I have done sampling to avoid repeating code. Once you understand a representative method, you can code on similar lines.

Listed below are few commonly used methods.

- `enterText(WebElement, String)`: This wrapper method enters text into textboxes or text areas. There are two arguments. One is of type `WebElement` and the other is of type String. The following code snippet shows the `enterText` method:

```
public void enterText(WebElement element,String text) {
        element.sendKeys(text);
}
```

- `clickElement(WebElement)`: This method clicks on the web element passed as an argument. The following code snippet shows the `clickElement` method:

```
public void clickElement(WebElement element) {
        element.click();
}
```

- `selectValue(WebElement, String)`: This method selects the text passed as an argument from the options of a `Select` object passed as the first argument. The declaration of the `Select` object should be done outside the method. The following code snippet shows the `selectValue` method:

```
Select select = null;
public void selectValue(WebElement element, String text) {
  select = new Select(element);
      select.selectByVisibleText(text);
}
```

- `getElementText(WebElement)`: This method returns the visible text (not hidden by CSS) of the element passed as a parameter. The following code snippet shows the `getElementText` method:

```
public String getElementText(WebElement element) {
    return element.getText();
}
```

- `isElementDisplayed(WebElement)`: This method returns true if the `WebElement` passed as a parameter is displayed on the screen. Notice that the method returns a Boolean. The following code snippet shows the `isElementDisplayed` method:

```
public boolean isElementDisplayed(WebElement element) {
    return element.isDisplayed();
}
```

The difference between quit() and close()

Two very important methods available in the WebDriver interface are `quit()` and `close()`. For beginners in Selenium, this is a topic of confusion because of a lack of understanding of its usage. Let's next understand both these methods, using these distinctions:

- `driver.quit()`: The `quit()` method quits the driver, closing every associated window.
- `driver.close()`: The `close()` method closes the currently focused window, quitting the driver if the current window is the only open window. If there are no windows open, it will error out.

We will be looking in detail at these two methods when we learn about handling popups.

Understanding the keyword driven framework

Next, we come to a very important juncture, where we come to understand what a Keyword-Driven framework is. A Keyword Driven framework is one in which keywords are written in an external file, such as .xls or xlsx . The common practice is to use the Keyword-Driven framework along with test data. Such a framework is called a Hybrid Keyword Framework.

We will have a very simple Excel document for this framework, which will be called Framework.xls, and this sheet will have two tabs, namely TestConfig and TestCases. We have already seen in Chapter 2, *Understanding the Document Object Model and Creating Customized XPaths*, what the TestConfig tab looks like, and we also extracted records for which the Execute column was marked as *Y*. Let's see what the TestCases tab looks like.

The following screenshot shows the TestCases tab:

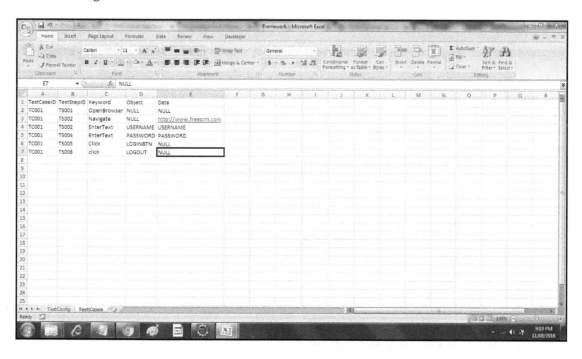

The TestCaseID in the TestCases tab should map to the TestConfig tab. If the Execute flag for test case TC001 is 'Y' in the TestConfig tab, then all the test steps for test case TC001 should execute in order from top to bottom.

Let's first extract this data by modifying our initial query which was as shown here:

```
select TestCaseID from TestConfig where Execute_Flag='Y'
```

The modified query is shown here:

```
Select * from TestCases where TestCaseID in (select TestCaseID from
TestConfig where Execute_Flag='Y')
```

Let's extract data from the `TestCases` tab now.

Let's have a look at an Extract Program.

The code shown below uses the previous query to extract `TestCases` steps based on Execute Flag:

```java
    public static void main(String[] args) throws FilloException {
        Fillo fillo = new Fillo();
        Connection conn = null;
        Recordset rcrdset = null;
        try {
            conn = fillo
.getConnection("C:\\SeleniumWD\\src\\main\\resources\\Framework.xlsx");
            String query = "Select * from TestCases where TestCaseID in (select
TestCaseID from TestConfig where Execute_Flag='Y')";

            rcrdset = conn.executeQuery(query);

            while (rcrdset.next()) {
              System.out.println(rcrdset.getField("TestCaseID") + ":::"
                + rcrdset.getField("Keyword") + ":::"
                + rcrdset.getField("Object") + ":::"
                + rcrdset.getField("Data"));
            }
        } catch (FilloException e) {
            throw new FilloException("Error in query");
        } finally {
            rcrdset.close();
            conn.close();
        }
    }
```

The output from the previous code is as follows:

```
TC001:::OpenBrowser:::NULL:::NULL
TC001:::Navigate:::NULL:::http://www.freecrm.com
TC001:::EnterText:::USERNAME:::USERNAME
TC001:::EnterText:::PASSWORD:::PASSWORD
```

```
TC001:::Click:::LOGINBTN:::NULL
TC001:::click:::LOGOUT:::NULL
```

One important thing to be noted about this output is that, after getting it, we need to send it to the framework so that the UI automation can proceed. For this purpose, we will require an `ArrayList` of Map, which we will see in the next section. Let's first look at how we extract the test data.

The data in bold is extracted from the `Data` column of the Excel sheet. If you notice, for the Navigate keyword we have the data hardcoded, but the data in bold seems to be in the form of variables of some sort. And, of course, they are variables. They represent what are called the `Keys` of a property file—say, for example, the name of the property file is `Login.properties`, the contents of which are shown here:

USERNAME=admin
PASSWORD=adminpass

The value on the left side of = is the key and the one on the right is the value.

The code to extract the values of the corresponding `Keys` is given here:

```
public static void main(String[] args) throws IOException {
  Properties prop = new Properties();
  InputStream input = null;
  try {
    input = new FileInputStream(
        "C:\\SeleniumWD\\src\\main\\resources\\Login.properties");
    prop.load(input);
    System.out.println("Username: " + prop.getProperty("USERNAME"));
    System.out.println("Password: " + prop.getProperty("PASSWORD"));
  } catch (FileNotFoundException e) {
    throw new FileNotFoundException();
  } finally {
    input.close();
  }
}
}


The output from the above program is
Username: admin
Password: adminpass
```

Concept of Map and HashMap

Map is an interface in Java that works on a **Key and Value pair** concept, which cannot contain duplicate keys and each Key maps to, at the most, one value. HashMap is the implementation class of Map, which we will be using to store the data from the Excel sheet, which will be related to test steps of a particular testcase. First of all, we will just require a Java HashMap to extract the test steps. One more testcase has been added to show what happens when only a HashMap is used to extract multiple test case steps.

The following code shows how to extract data in HashMap from the `Framework.xls`:

```
public static void main(String[] args) {
  Fillo fillo = new Fillo();
  Map<String, String> testcaseData = new HashMap<String, String>();
  Connection conn = null;
  Recordset rcrdset = null;
  try {
    conn = fillo
.getConnection("C:\\SeleniumWD\\src\\main\\resources\\Framework.xlsx");
    String query = "Select * from TestCases where TestCaseID in (select
TestCaseID from TestConfig where Execute_Flag='Y')";
    rcrdset = conn.executeQuery(query);
    while (rcrdset.next()) {
      testcaseData.put("TestCaseID", rcrdset.getField("TestCaseID"));
      testcaseData.put("Keyword", rcrdset.getField("Keyword"));
      testcaseData.put("Object", rcrdset.getField("Object"));
      testcaseData.put("Data", rcrdset.getField("Data"));
    }
    System.out.println(testcaseData);
  } catch (FilloException e) {
    throw new FilloException("Error in query processing");
  } finally {
    rcrdset.close();
    conn.close();
  }
}
}
```

```
The output of this program is
{Keyword=click, TestCaseID=TC002, Object=LOGOUT, Data=NULL}
```

As seen from the preceding output, the last test step of the test case `TC002` is fetched into the HashMap. This happens because a HashMap does not allow duplicate values in the Key and it overwrites the last value every time a duplicate Key appears. Hence, we get only the last test step, since all initial steps were overwritten.

To overcome this issue, we require an `ArrayList` of HashMaps, which can be done using the following code:

```
public class ExtractAllData {

  public static void main(String[] args) throws FilloException {
  Fillo fillo = new Fillo();
  Map<String, String> testcaseData = new HashMap<String, String>();
  List<Map> teststepList = new ArrayList<Map>();
  Connection conn = null;
  try {
  conn = fillo
  .getConnection("C:\\SeleniumWD\\src\\main\\resources\\Framework.xlsx");
  String query = "Select * from TestCases where TestCaseID in (select
  TestCaseID from TestConfig where Execute_Flag='Y')";
  Recordset rcrdset = null;
  rcrdset = conn.executeQuery(query);
  while (rcrdset.next()) {
  testcaseData = new HashMap<String, String>();
  testcaseData.put("TestCaseID", rcrdset.getField("TestCaseID"));
  testcaseData.put("Keyword", rcrdset.getField("Keyword"));
  testcaseData.put("Object", rcrdset.getField("Object"));
  testcaseData.put("Data", rcrdset.getField("Data"));
  teststepList.add(testcaseData);
  }
  System.out.println("List Content size: " + teststepList.size());
  System.out.println("List Contents: " + teststepList);
  rcrdset.close();
  conn.close();
  } catch (FilloException e) {
  throw new FilloException("Error in query");
  }
  }
}
```

Do not forget to put this line of code in the `recordSet` iteration:
```
        testcaseData = new HashMap<String, String>();
```
Without this line, the `teststepList` will contain 12 duplicated entries of the last test step, which is not correct.

When the preceding code executes, we will have all the test steps that have to be executed. This completes the extraction process.

Bird's eye view of the framework

We will now learn how our Keyword-driven framework will be structured from a high level.

The following figure shows the bird's eye view of the framework:

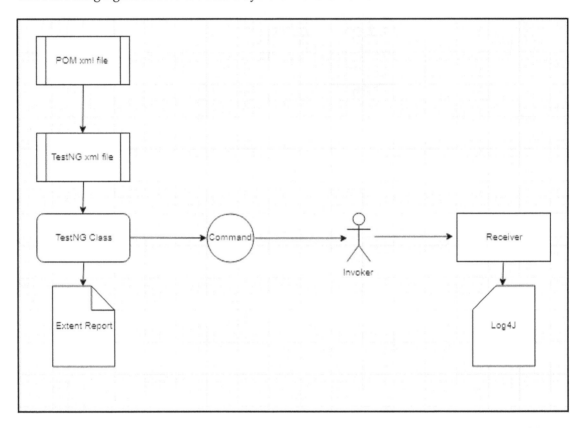

We will be using a TestNG framework to control all the components of our framework. TestNG is a framework that is a combination of NUnit and JUnit frameworks. This is our base framework. The testng.xml is triggered from pom.xml, which eventually will be executed from Jenkins.

Introducing the testng XML file

The framework will be controlled by TestNG XML, which is an XML file with all the classes that have to be executed as part of the test run.

The `TestNG.xml` is shown in the following code and `DriverScript` will be the name of our main class:

```
<!DOCTYPE suite SYSTEM "http://testng.org/testng-1.0.dtd" >
<suite name="Suite1" verbose="1" >
    <test name="FirstTest">
        <classes>
            <class name="packt.keywordframework.DriverScript"/>
        </classes>
    </test>
</suite>
```

Even though this will be the first class to be executed, it will not contain a Main method. We will see how we can trigger this class in a `Chapter 7`, The Command Pattern and Creating Components.

Triggering the testng XML from within the POM XML file

The `testng.xml` that is shown in the preceding code will be triggered from `pom.xml`. Let's understand how this triggering takes place. For the triggering to happen, a plugin called `Maven Surefire Plug-in` is used. The corresponding dependency is placed in the `pom.xml` inside the `<plugin></plugin>` tag, as shown in the following code snippet.

Apart from the Maven Surefire plugin, another plugin called the Maven Compiler plugin is also required. For integration tests, the Maven failsafe plugin should be used.

The following snippet of the `pom.xml` shows the `<build></build>` tags, inside which are `<plugins></plugins>` tags. Place the dependencies inside the plugins tag, as shown in the following code (make sure to select Version 1.8 of source and target):

```
<build>
 <plugins>
    <plugin>
        <groupId>org.apache.maven.plugins</groupId>
        <artifactId>maven-surefire-plugin</artifactId>
        <version>2.22.0</version>
        <configuration>
```

```
            <suiteXmlFiles>
                <suiteXmlFile>
                        testng.xml
                </suiteXmlFile>
            </suiteXmlFiles>
        </configuration>
    </plugin>

    <plugin>
        <groupId>org.apache.maven.plugins</groupId>
        <artifactId>maven-compiler-plugin</artifactId>
        <version>3.8.0</version>
        <configuration>
            <source>1.8</source>
            <target>1.8</target>
        </configuration>
    </plugin>
  </plugins>
 </build>
```

After adding this dependency, save the project and right click on the **Project** in Eclipse. Click **Maven | Update Project**.

The next thing to do is right click on pom.xml and select **Run As | Maven Test**.

Alternatively, you can run Maven from the command line using the command mvn clean verify (using the Maven failsafe plugin) or mvn clean test (using the surefire plugin). JDK that is greater than 1.8 is required. The .java files in the project will be compiled and the tests in the testng.xml will be run. Right now, we only have one class DriverScript.java in testng.xml. Since we have not created this class yet, an error will be thrown after executing the previous command. But that's not a problem.

Handling errors while running pom.xml

If an error that's stating 'no compiler is provided in this environment. perhaps you are running on a jre rather than a jdk?' is displayed, then these steps have to be performed:

1. Right-click on the **Project**. Select Properties and go to the **Compiler** section.

2. Verify that the compiler compliance level is set at 1.8 since we are using Java 8, as shown in the following screenshot:

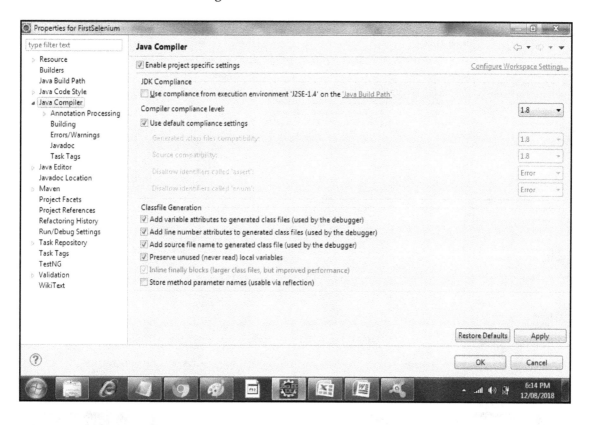

3. Go to **Window | Preferences** and verify that the JDK 1.8 checkbox is checked, as shown in the following screenshot:

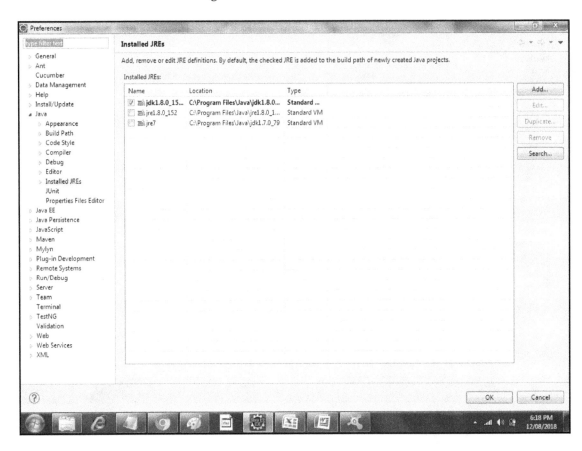

Once these settings are done, the aforementioned error will no longer appear.

Introducing the log4j framework

Debugging the framework is an equally important activity when errors occur. Logging is the process of creating logs of the framework execution. Because of this, errors occurring during the execution can be easily diagnosed, since we know from the log about the last successful action that happened.

Log4j is an open source framework for logging. For using the Log4J framework, the following dependency has to be added in `pom.xml`:

```xml
<!-- https://mvnrepository.com/artifact/log4j/log4j -->
<dependency>
 <groupId>log4j</groupId>
 <artifactId>log4j</artifactId>
 <version>1.2.17</version>
 <scope>test</scope>
</dependency>
```

Follow these steps to download the Log4J dependency:

1. Save the `pom.xml`
2. Right-click on **Project** and select **Update Project**
3. The required dependency will get downloaded

In the case of firewall restrictions, a manual download of the log4J JAR has to be done, and later this Jar has to be added to the project build path by right-clicking on the project in Eclipse, then selecting properties. In the properties popup, select Java Build Path and click on Libraries. Add the log4j jar by clicking Add External JARs and selecting the JAR from the local drive.

For using Log4J, we either require a `log4j.xml` or `log4j.properties` for configuring log4j to generate logs. We will be using `log4j.properties` in this chapter.

`Log4J.properties` is shown in the following snippet:

```
#Define the root logger. Level of Root logger is defined as DEBUG
log4j.rootLogger=DEBUG, CONSOLE, LOGFILE

#Define the CONSOLE Appender, the threshhold, the layout and the conversion
#pattern
# The logging levels are in the order ALL(Integer.MAX_VALUE) < TRACE(600) <
#DEBUG(500) < #INFO(400) < WARN(300) < ERROR(200) < FATAL(100) < OFF(0).
Since we #have kept the logging #level # at INFO here, all
INFO,WARN,ERROR,FATAL messages #will be displayed. The DEBUG and #TRACE
level messages are not displayed

log4j.appender.CONSOLE=org.apache.log4j.ConsoleAppender
log4j.appender.CONSOLE.Threshold=INFO
log4j.appender.CONSOLE.layout=org.apache.log4j.PatternLayout
log4j.appender.CONSOLE.layout.ConversionPattern=%d{yyyy-MM-dd HH:mm:ss}
%-5p %C{1}:%L - %m%n
# Define the File Appender
log4j.appender.LOGFILE=org.apache.log4j.RollingFileAppender
```

```
log4j.appender.LOGFILE.File=D:\\logging.log
log4j.appender.LOGFILE.MaxFileSize=20MB
log4j.appender.LOGFILE.MaxBackupIndex=5
log4j.appender.LOGFILE.Append=true
log4j.appender.LOGFILE.Threshold=INFO
```

The log messages will be pushed to the `D:\logging.log` file, which will be appended. On reaching 50 MB, a new file will get created. The new file will be an archived version and in the type it will have the version number listed. For example, the first file will be of type `1`, the second of type `2`. When the version 2 file is created, the version 1 file data will be pushed to version 2 and the current version data will be pushed to version 1. The next thing worth mentioning is that there can be only five versions that can be created. After the fifth version, once `maxFileSize` is reached, data will start flowing from the current version to version 1, from version 1 to version 2, version 2 to version 3, and so on. Data in version 5 that was originally there will be overwritten by data from version 4.

Here is what a sample log looks like:

```
2018-08-14 14:29:05 INFO Navigate:101 - Navigating to
http://www.freecrm.com
2018-08-14 14:29:11 INFO WorkspaceSelection:35 - Created Logger object
2018-08-14 14:29:11 INFO WorkspaceSelection:90 - Entering
SelectWindowByTitleAndURL
2018-08-14 14:29:16 INFO WorkspaceSelection:98 - windowset size: 2
```

Let's dissect a sample line from the log:

```
2018-08-14 14:29:16 INFO WorkspaceSelection:98 - windowset size: 2
```

The lines from `log4j.properties` that correspond to the following output are as follows:

```
log4j.appender.LOGFILE.layout=org.apache.log4j.PatternLayout
 log4j.appender.LOGFILE.layout.ConversionPattern=%d{yyyy-MM-dd HH:mm:ss}
%-5p %C{1}:%L - %m%n
```

In the previous two lines of code, `%d{yyyy-MM-dd HH:mm:ss}` corresponds to `2018-08-14 14:29:16` in the output. `%-5p` corresponds to the text `INFO`. The `'-'` in `%-5p` indicates that the entry should be left justified with maximum five characters. `'%C{1}:'` corresponds to `'WorkspaceSelection:'`. This specifies the class name. `'%L -'` corresponds to the line number in the class followed by a hyphen. `'%m` corresponds to the application supplied message `windowset size: 2`. `%n` corresponds to the platform dependent line separator character or characters.

For details on the Log4J framework, refer to `https://logging.apache.org/log4j/2.x/javadoc.html`.

Extent Reports

Extent Reports are graphical reports, which are a graphical presentation of the Automation Run.

The following screenshot is a sample extent report. A tidy report displaying the number of tests passed is displayed:

To use Extent Reports, add the following dependency in pom.xml, as shown here:

```
<!-- https://mvnrepository.com/artifact/com.aventstack/extentreports -->
<dependency>
 <groupId>com.aventstack</groupId>
 <artifactId>extentreports</artifactId>
 <version>3.1.5</version>
 <scope>test</scope>
</dependency>
```

Follow the following steps to download the Extent Report Jar:

1. Save the `pom.xml`
2. Right click on **Project**
3. Click **Maven | Update Project**

Alternatively, the Extent Report jar can be downloaded from `http://extentreports.com/community/` (click on the **Java 3.1.5** link).

How to use Extent Reports in code?

To use Extent Reports in code, follow the following steps:

1. Create `ExtentReports` and `ExtentTest` objects as `extentReport` and `extentTest` respectively.
2. Instantiate the `ExtentReports` object: `extentReport = new ExtentReports (System.getProperty("user.dir") +"/test-output/SampleReport.html", true);`.
3. Add other details and load the configuration file. The config file is optional:

```
extentReport
.addSystemInfo("Host Name", "SampleHostName")
.addSystemInfo("Environment", "Demo")
.addSystemInfo("User Name", "Pinakin Chaubal");
extent.loadConfig(new
File(System.getProperty("user.dir")+"\\extent-config.xml"));.
```

4. Start the test using `extentTest = extentReport.startTest("Test1");`.
5. Log the result using `extentTest.log(LogStatus.PASS, "Test Case Create Invoice has Passed");`.
6. At the end of the test, insert `extentReport.endTest(extentTest)`.
7. At the end of the suite, insert `extentReport.flush();` and `extentReport.close();`. Only after the flush command will data appear in the report. These should be in the final block.

There are eight statuses that can be output to the report: Error, Fail, Fatal, Info, Pass, Skip, Unknown, and Warning.

Summary

This chapter explored basic Selenium commands. We learned what method signatures are and came to understand the concept of List in java. We also learnt the frequently used methods in Selenium and created some reusable methods. The basic difference between `quit()` and `close()` was covered as well. We then had a look at what a Keyword-driven Framework was and what Maps are. The framework that we are going to design in the upcoming chapters was looked at from a bird's eye view. From this, we learned what Log4J and ExtentReports are.

In the next chapter, we will take a small diversion and look at two of the important concepts in Framework design: "Popups" and "Alerts". We will come to understand the concept of window handle. We will look at sets and arrays of window handles. We will write code to enter data in a pop-up form. In this process, we will also create a pop-up form if needed. So, stay tuned.

4
Handling Popups, Frames, and Alerts

This is one of the most important chapters, focusing on handling popups and alerts. Every website nowadays has some sort of popup and in this chapter, we will see what popups and alerts are and how to handle them.

The following topics will be covered in this chapter:

- Window handles
- Modal and non-modal dialogs
- Understanding the set interface
- Getting to know iterators
- A look at pop-up windows and alerts
- Creating a demo page for popups and alerts
- A look at frames and iframes
- Mobile popups and alerts

Technical requirements

You will be required to have Notepad++ in addition to the software mentioned in `Chapter 3`, *Basic Selenium Commands and Their Usage in Building a Framework*.

The code files of this chapter can be found on GitHub:
`https://github.com/PacktPublishing/Selenium-WebDriver-Quick-Start-Guide/tree/master/Chapter04`

Check out the following video to see the code in action:
`http://bit.ly/2CGsWbi`

Window handles

Let's understand what a window handle is. A window handle is an alphanumeric ID that gets assigned to each window when the object of WebDriver is instantiated. Selenium uses this ID to identify each window and switch between them.

Fetching the window handles

In order to switch between all available windows, Selenium first needs to get hold of the window handles. For this purpose, there are two methods provided: `getWindowHandle()` and `getWindowHandles()`:

- `getWindowHandle()`: This method, when invoked on the WebDriver object, returns the handle of the current window which has the focus
- `getWindowHandles()`: This method, when invoked on the WebDriver object, returns a set of all open windows

Understanding the Set interface

Set is an interface that extends the collection interface. Duplicate values cannot be contained in a set. The values in a set are not stored in an ordered manner and hence these values cannot be accessed using an index. There are three different classes that implement the Set interface:

- `HashSet`:
    - Stores elements by using hashing a mechanism
    - An index cannot be used to access values
    - Can contain unique elements only
    - `LinkedHashSet`
    - Maintains the insertion order
    - `TreeSet`
    - Maintains ascending order

There are methods available in the Set interface, such as `add()`, `contains()`, `isEmpty()`, `iterator()`, `remove()`, `size()`, and so on, out of which we will be looking at an iterator.

A look at the iterator() method

Each of the classes in the Collection framework provide an `iterator()` method that returns an iterator object to the start of the collection. The iterator object implements either the Iterator or the `ListIterator` interface. There are two main differences between the Iterator and `ListIterator` interfaces:

- Iterator is used for traversing both list and set. `ListIterator` is used to traverse only lists.
- Only forward traversal is possible using Iterator. Bi-directional traversal is possible using `ListIterator`.

Modal and non-modal dialog

The two types of dialog boxes encountered in a web app are modal and non-modal dialog boxes:

- A modal dialog box is one which disables the main content until you interact with the modal dialog.
- A non-modal dialog box, on the other hand, allows you to interact with the main content even when the non-modal dialog box is present.

We will be looking at two subcategories of modal and non-modal dialog: pop-up windows and JavaScript alerts. We will be looking at scenarios covering modal and non-modal dialog for both popups and alerts.

Modal and non-modal pop-up windows

A separate window that shows up on the screen when a link or button is clicked is known as a pop-up window. Pop-up windows can be modal and non-modal. When a popup is non-modal, it opens up as a regular window and when the popup is modal, it usually shows up within a frame in the parent window. It may also show up inside the main HTML content and not specifically in a frame.

JavaScript and jQuery alerts

JavaScript alerts are ones that are usually modal. One has to do some action on the alert before the content in the parent window can be accessed. These alerts can have **OK** (**OK** and **CANCEL**) or (**YES** and **NO**) buttons. There is usually some message text displayed on the alert.

jQuery alerts can be designed to be non-modal by implementing the fade-in feature of jQuery. Such an alert appears for a short duration and then slowly fades away.

We will be creating both JavaScript modal and jQuery non-modal popups (using the jQuery fade-in feature).

Handling non-modal popup windows

We will be using the website http://popuptest.com/goodpopups.html to demonstrate non-modal pop-up windows. We will write a small program to load this URL and then click the **Good PopUp #1 link**. A pop-up window will open and then we will fetch the window handles of both the open windows.

The following code clicks a link on the home page and fetches the window handles of both the open windows as well as their titles:

```
WebDriver driver = new ChromeDriver();
driver.manage().window().maximize();
driver.navigate().to("http://popuptest.com/goodpopups.html");
driver.findElement(By.xpath("//a[text()='Good PopUp #1']")).click();
Set openWindows = driver.getWindowHandles();
System.out.println("No of open windows: " + openWindows.size());
Iterator<String> it = openWindows.iterator();
String parent = it.next();
System.out.println("Parent Window: " + parent + " title: " +
driver.getTitle());
String child = it.next();
System.out.println("Child Window:" + child + " title: "
 + driver.getTitle());
--------------------------------------------------------------------
----

Output:
No of open windows: 2
Parent Window: CDwindow-0F9E3C139C7C9BFEE7D9B981F83425D6 title: PopupTest -
test your popup killer software
```

```
Child Window:CDwindow-1BD6347CF08C0721D66F726F9D04F52C title: PopupTest -
test your popup killer software
```

Did you notice something funny about the output? We have two windows and two unique window handles but the title printed in both the windows is the same. How can that be possible? This happened because we did not switch windows before calling the getTitle() method. Here comes the concept of switching windows.

Introducing the switchTo() method

Selenium has to be explicitly told which window to focus on. In this case, it has focused on the first parent window. We switch windows in Selenium using the switchTo() command, which is invoked on the driver object. The signature of switchTo is driver.switchTo.window(String). The switchTo() method is in the WebDriver interface, which returns an object of the TargetLocator interface: a child interface of WebDriver. On this object of TargetLocator, the window method is invoked and this returns a WebDriver object. The parameter is of a string type, which is the alpha-numeric handle of the window. Let's handle the small issue in our code by adding switchTo().window(String) at two places in our code right after extracting the window handles for parent and child windows. The string parameters will be parent and child.

The output After this addition is shown here:

```
No of open windows: 2
Parent Window: CDwindow-3378E656C226EBA5B5880A0DE93D1E24title: PopupTest -
test your popup killer software
Child Window:CDwindow-76B76D2A879C440BBC8A548230320459title: PopupTest
Sunday August, 19 2018
```

Looking at the getTitle() method

We can perform operations on a particular window by using the switchTo() command first and then verifying the title. If the title of the window matches the one we want, we can start performing operations on that window or we can invoke the switchTo method again. We can iterate through all open windows using the while loop. Apart from this, we can also search for the existence of elements, we can fire up JavaScript events to check that some JavaScript function exists, and so on.

In the condition of the while loop, we utilize the `hasNext()` method on the iterator object. The `hasNext` method keeps on fetching values in the iterator object until it reaches the last value.

The following piece of code can be used to switch to the desired window. The code issues a break after finding the window with the desired title:

```
public void switchWindowByTitle(String expectedTitle){
    Set<String> windowHandles = driver.getWindowHandles();
  Iterator it = windowHandles.iterator();
  while (it.hasNext()){
        driver.switchTo(it.next());
        if (driver.getTitle().equalsIgnoreCase(expectedTitle)){
            break;
        }
    }
}
```

Looping through all open windows using the simple for loop

Sometimes, using a simple for loop instead of a set can prove easy to comprehend. Let's look at the `ArrayList` method with the same link `http://popuptest.com/goodpopups.html`.

The following code, when run from a main method, prints the titles of each popup:

```
WebDriver driver = new ChromeDriver();
driver.manage().window().maximize();
driver.navigate().to("http://popuptest.com/goodpopups.html");
try {
} catch (InterruptedException e) {
// TODO Auto-generated catch block
e.printStackTrace();
}
driver.findElement(By.xpath("//*[text()='Good PopUp #1']")).click();
for (String i : driver.getWindowHandles()) {
driver.switchTo().window(i);
System.out.println(driver.getTitle());
}
}
```

Handling modal popup windows

In this section, we will be using a simple modal popup created using `http://www.w3schools.org/` and then we will write a script to handle that popup. The simple modal pop-up code is given next. We will build the HTML page first and then we will plug the CSS and JavaScript. We have kept the CSS and JavaScript external for the sake of simplicity:

```html
<html>
  <head>
    <link rel="stylesheet" type="text/css" href="GetModal.css">
  </head>
  <body onload="test()">
    <!-- Trigger/Open The Modal -->
    <button id="myBtn">Open Modal</button>
    <!-- The Modal -->
    <div id="myModal" class="modal">
    <!-- Modal content -->
    <frame name="MainWindow">
        <div class="modal-content">
        <div class="modal-header">
        <span class="close">&times;</span>
        <h2>Login Form</h2>
    </div>
    <div class="modal-body">
        <table id="credentials">
        <tr>
            <td>UserID</td>
            <td><input type="text" id="userid"/></td>
        </tr>
        <tr>
            <td>Password</td>
            <td><input type="text" id="password"/></td>
        </tr>
        <tr>
            <td><input type="button" id="ok" value="Ok"></td>
            <td><input type="button" id="cancel" value="Cancel"></td>
        </tr>
        </table>
    </div>
    <div class="modal-footer">
        <h3>Login to Application</h3>
    </div>
  </div>
  </frame>
  </div>
  </body>
```

```
<script src="GetModal.js"></script>
</html>
```

Next, we will look at the CSS. The CSS is used to give styling to our HTML. So let's look at the CSS for this HTML:

```
/* The Modal (background) */
.modal {
    display: none; /* Hidden by default */
    position: fixed; /* Stay in place */
    z-index: 1; /* Sit on top */
    left: 0;
    top: 0;
    width: 100%; /* Full width */
    height: 100%; /* Full height */
    overflow: auto; /* Enable scroll if needed */
    background-color: rgb(0,0,0); /* Fallback color */
    background-color: rgba(0,0,0,0.4); /* Black w/ opacity */
}
/* Modal Content/Box */
.modal-content {
    background-color: #fefefe;
    margin: 15% auto; /* 15% from the top and centered */
    padding: 20px;
    border: 1px solid #888;
    width: 80%; /* Could be more or less, depending on screen
    size */
}
/* The Close Button */
.close {
    color: #aaa;
    float: right;
    font-size: 28px;
    font-weight: bold;
}
.close:hover,
.close:focus {
    color: black;
    text-decoration: none;
    cursor: pointer;
}
/* Modal Header */
.modal-header {
    padding: 2px 16px;
    background-color: #5cb85c;
    color: white;
}
/* Modal Body */
```

```
.modal-body
{
    padding: 2px 16px;
}
/* Modal Footer */
.modal-footer {
    padding: 2px 16px;
    background-color: #5cb85c;
    color: white;
}
/* Modal Content */
.modal-content {
    position: relative;
    background-color: #fefefe;
    margin: auto;
    padding: 0;
    border: 1px solid #888;
    width: 80%;
    box-shadow: 0 4px 8px 0 rgba(0,0,0,0.2),0 6px 20px 0
                rgba(0,0,0,0.19);
    animation-name: animatetop;
    animation-duration: 0.4s
}
/* Add Animation */
@keyframes animatetop {
    from {top: -300px; opacity: 0}
    to {top: 0; opacity: 1}
}
```

Now comes the final part of plugging in the JavaScript. Plug this in the HTML right before the ending script tag. The file should be named ModalPopup.html:

```
 function test(){
// Get the modal
var modal = document.getElementById('myModal');
// Get the button that opens the modal
var btn = document.getElementById("myBtn");
// Get the <span> element that closes the modal
var span = document.getElementsByClassName("close")[0];
// When the user clicks on the button, open the modal
btn.onclick = function() {
modal.style.display = "block";
}
// When the user clicks on <span> (x), close the modal
span.onclick = function() {
modal.style.display = "none";
}
// When the user clicks anywhere outside of the modal, close it
```

```
window.onclick = function(event) {
if (event.target == modal) {
modal.style.display = "none";
 }
 }
}
```

All the pieces are separate from each other. It becomes an HTML page created using plain HTML with inclusions for CSS and JavaScript. Clicking the cross closes the popup. The page does not have functionality for the **OK** and **Cancel** buttons. It is just a dummy page for demo purposes.

The following code opens up the demo page, clicks on the button, and then a modal popup opens, after which a username and password are entered:

```
WebDriver driver = new ChromeDriver();
// driver.manage().window().maximize();
driver.navigate().to(
"C:\\Users\\pinakin.chaubal\\Desktop\\ModalPopup.html");
driver.findElement(By.id("myBtn")).click();
System.out.println(driver.findElement(By.tagName("body")).getText());
driver.findElement(By.id("userid")).sendKeys("testuser");
driver.findElement(By.id("password")).sendKeys("testpass");
```

Modal and non-modal alerts

Next, we will see how to create modal and non-modal alerts and also handle those through Selenium code. In modal alerts, we will see JavaScript alerts, and in non-modal alerts, we will see jQuery alerts. Let's first create some alerts.

Creating JavaScript and jQuery alerts

The following is the code to generate a simple modal alert in JavaScript. This alert in the code will appear on the main page and one cannot access the main page in the background until either the **OK** or **Cancel** button is clicked:

```
<html>
 <head>
    <script type="text/javascript">
       function display_alerts(){
          var text1;
      if (confirm("Click")) {
       text1 = "Ok was pressed";
```

```
          } else {
           text1 = "Cancel was pressed";
          }
          document.getElementById("disptext").innerHTML = text1;
            }
        </script>
    </head>
    <body>
        <input type = "text" id = "text1"/>
        <input type = "button" onclick="display_alerts()" value="Display
        Alert" />
      <p id="disptext"/>
      </body>
</html>
```

The following is the code to generate a simple non-modal alert in jQuery. This alert appears
for a certain amount of time and then disappears. The method of disappearance can be
configured. Here we have configured a fade-away alert:

```
<html>
  <head>
    <style>
        .message {
            position: absolute;
            top: 145px;
            width: 60%;
            left: 20%;
            background: #ff9;
            border: 1px solid #fc0;
            font-size: 150%;
            padding: 15px 50px;
        }
        .message p {
            margin: 0;
            padding: 0 40px 0 20px;
            text-align: center;
}
    </style>
    </head>
    <body>
        <div class="message">
            <p>This is a non-modal alert</p>
        </div>
        <script src="C:\Users\Bhagyashree\Documents\jquery-
        3.3.1.min.js">                         </script>
        <script>
            $(document).ready(function(e) {
                // Check if there's a message
```

```
                        if($('.message').length) {
                        // If mouse is clicked, moved or a key is pressed
                        $(document).one('click mousemove keypress',
                         function(e) {
                        // Fade the message away after 500 ms
                        $('.message').animate({ opacity: 1.0 },
                        500).fadeOut();
                        });
                }
            });
        </script>
    </body>
</html>
```

Handling non-modal jQuery alerts

When we created the previous jQuery alerts, we did it with the fadeout functionality. When we invoke the fadeOut() function of jQuery, the alert remains until the timeout value is specified and then it is slowly faded away. In such a situation, we don't need to explicitly handle the popup and that is the reason it is considered a non-modal alert. You may try this out.

Handling modal JavaScript alerts

We have created a sample HTML page with a textbox and button. When the **Display Alert** button is clicked, the modal alert is displayed with **OK** and **Cancel** buttons. At this point, the textbox on the main page cannot be accessed. When the **OK** or **Cancel** buttons are clicked, an appropriate message is displayed on the main page.

The following code will click on the **Display Alert** button and later will click once on the **Ok** button of the modal alert, and the next time will click on the **Cancel** button:

```
WebDriver driver = new ChromeDriver();
driver.navigate().to(
"C:\\modalalert.html");
  driver.findElement(By.id("dispalert")).click();
Alert alert = driver.switchTo().alert();
alert.accept();
driver.findElement(By.id("text1")).sendKeys("Ok clicked");
driver.findElement(By.id("dispalert")).click();
driver.switchTo().alert();
alert.dismiss();
```

```
driver.findElement(By.id("text1")).clear();
driver.findElement(By.id("text1")).sendKeys("Cancel clicked");
```

In the preceding code, we find two new methods invoked on the Alert object, which are `accept()` and `dismiss()`:

- `accept()`: Clicks on the **Ok** button of the alert
- `dismiss()`: Clicks on the **Cancel** button of the alert

These methods will throw a `NoSuchAlertError` if the alert does not exist at that time.

Understanding frames and iframes

Frames are HTML pages that are used inside the main HTML tags. A frame is an HTML tag written as `<frame>`. It is used to divide the web page into various sections. It specifies each frame within the `<frameset>` tag. Iframes or Inline frames is written as `<iframe>`and this is also another tag in HTML, but it is used to embed some other document into the current HTML document. The method to handle frames and iframes in Selenium is the same. We will only see how to handle frames here.

Refer to the `FramesDemo.html` given in the following code:

```
<!DOCTYPE html>
<html>
<frameset cols="25%,50%,*">

  <frame name="frame1" src="C:\Frame1.html">
  <frame name="frame2" src="C:\Frame2.html">
  <frame name="frame3" src="C:\Frame3.html">
</frameset>
</html>
```

The code for `Frame1.html` is given in the following code:

```
<!DOCTYPE html>
<html>
 <body>
 <h1>Frame 1</h1>
 <p>Contents of Frame 1</p>
 <p><input type="radio" id="radio1"/></p>
 </body>
</html>
```

The code for `Frame2.html` is given in the following code:

```
<!DOCTYPE html>
<html>
 <body>
 <h1>Frame 2</h1>
 <p>Contents of Frame 2</p>
 <p><input type="text" id="input1"/></p>
 </body>
</html>
```

The code for `Frame3.html` is given in the following code:

```
<!DOCTYPE html>
<html>
 <body>
 <h1>Frame 3</h1>
 <p>Contents of Frame 3</p>
 <p><input type="checkbox" id="chk1"/></p>
 </body>
</html>
```

The following is the code to handle elements in three frames:

```
WebDriver driver = new ChromeDriver();
driver.navigate().to("C:\\FramesDemo.html");
 driver.switchTo().frame(0);
 System.out.println(driver.findElement(By.tagName("body")).getText());
 driver.findElement(By.id("radio1")).click();
 driver.switchTo().defaultContent();
 driver.switchTo().frame("frame2");
 driver.findElement(By.id("input1")).sendKeys("input box");
 driver.switchTo().defaultContent();
 driver.switchTo().frame("frame3");
 driver.findElement(By.id("chk1")).click();
```

In the preceding code, it can be seen that to click each of the web elements, we have to issue a command: `driver.switchTo.Frame(int)` or `driver.switchTo.Frame(String)`.

We have taken the examples of integer and string parameters. We pass the index as the integer parameter or the name of the frame as the string parameter. Whenever we issue this command, control goes to that particular frame and we can only access web elements inside that frame.

If there are nested frames inside a particular frame, the first navigation has to be to the container or parent frame followed by navigation to the child frame. If there are parallel frame hierarchies, as in the preceding example, one has to issue `driver.switchTo.defaultContent()` to get to the outermost frame if the control is currently in one particular frame and a web element in another parallel frame has to be clicked. If `driver.switchTo.defaultContent()` is not issued before accessing a parallel frame hierarchy, then a `NoSuchFrameException`, provided the control is in another frame at that moment. This exception is primarily thrown when a frame does not exist.

Handling mobile app permission alerts

Such a situation may arise quite a few times when we need to handle app permissions alerts. An app permissions dialog or alert looks like the one displayed in the following screenshot:

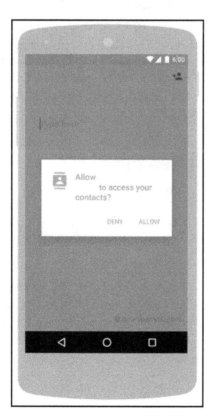

To understand how to handle the app permissions dialog on a mobile, understanding the DesiredCapabilities class is quite important. Let's understand the DesiredCapabilities class, after which we will see what Desired Capability has to be put in place for handling the app permissions dialog on Android in the *autograntpermissions capability in Android section*.

DesiredCapabilities

Let's next understand a very important class in Selenium: DesiredCapabilities. Before understanding the DesiredCapabilities class, knowing what Desired Capability means is necessary.

What is a Desired Capability?

 All server implementations will rarely support every feature of WebDriver. It is advised that the client and server should use JSON objects with certain predefined properties when describing which features a user requests that has to be supported by a session. If a session cannot support a certain capability that is requested in the Desired Capabilities, a capabilities object is returned, which is read-only, and this object indicates the capabilities the session supports.

The DesiredCapabilities class

DesiredCapabilities is a class in the org.openqa.selenium.remote.DesiredCapabilities package. It gives one the facility to set the properties of the browser, which are mainly BrowserName, Platform, and the version of the browser.

Generally, the DesiredCapabilities class is used when we need to use Selenium Grid and multiple test cases have to be executed on multiple systems with different browsers having different versions with different operating systems. Earlier, this was a major concern but now browsers autoupdate by themselves. We will see one example for Internet Explorer, where our tests may fail because of protected mode settings.

Enabling protected mode settings in Internet Explorer

Sometimes, while using Internet Explorer for the first time, a `Session not found` exception is thrown. To resolve this exception, do either of the following:

- Manually click the **Enable Protected Mode Setting** checkbox under the **Security** tab of **Internet Options**. Click the checkbox for **Internet, Local Intranet, Trusted Sites, and Restricted Sites**. This checkbox should be checked for all four tabs; the settings should be the same.
- Write the following code. The code ignores Internet Explorer's protected mode settings:

```
InternetExplorerOptions options = new InternetExplorerOptions();
options.IntroduceInstabilityByIgnoringProtectedModeSettings =
true;
WebDriver driver = new InternetExplorerDriver(options);
driver.get("http://www.google.com");
```

Here, Internet Explorer options are passed as an argument to the `InternetExplorerDriver` constructor. This code does the same work as is done by enabling the protected mode setting manually. We will look at Internet Explorer options in the next section.

`DesireCapabilities` is now going to be deprecated and hence Selenium has introduced Chrome, Firefox, and Internet Explorer options, which we will look at next.

Learning Chrome options

The WebDriver APIs provide techniques to pass capabilities to `ChromeDriver`. The `ChromeOptions` class is one such technique that is supported by Java.

Let's look at an example of the **Save password** dialog that shows up when you log in to an application on Google Chrome, as shown in the following screenshot:

We need to prevent this dialog from appearing, so we need to use ChromeOptions. We create an object of ChromeOptions, which is then passed to the ChromeDriver constructor. This object contains convenient methods for setting ChromeDriver-specific capabilities.

The following code disables the Save password popup for Chrome:

```
ChromeOptions options = new ChromeOptions();
Map<String, Object> prefs = new HashMap<String, Object>();
prefs.put("credentials_enable_service", false);
prefs.put("profile.password_manager_enabled", false);
options.addArguments("disable-extensions");
options.addArguments("--start-maximized");
options.setExperimentalOption("prefs", prefs);
System.out.println("Before logging");
logMessage(log, "Opening Browser");
System.out.println("After logging");
driver = new ChromeDriver(options);
```

Here, the log is the Logger object (Log4J). setExperimentalOption sets an experimental option. It is used for setting new ChromeDriver options that are not yet exposed through the ChromeOptions API. The preceding code starts a Chrome browser session in which notifications are suppressed.

Learning the Firefox options and the Firefox profile

We have similar options for the Firefox browser. The way to do that is by including the following two lines:

```
FirefoxOptions options = new FirefoxOptions();

WebDriver driver = new FirefoxDriver(options);
```

The Firefox profile is used because when we start the Firefox browser through Selenium using a temp profile with no plugins or bookmarks attached to it. If we want to start Firefox with either of these attached, then we need a Firefox profile. The following short code enables a flash player in Firefox:

```
FirefoxProfile ffprofile = new FirefoxProfile();
profile.setPreference('dom.ipc.plugins.enabled.libflashplayer.so',
'false'); FirefoxOptions ffoptions = new FirefoxOptions();
options.setProfile(ffprofile);
```

Learning the Internet Explorer options

We have similar options for the Internet Explorer browser. The way to do incorporate these options is by including the two lines. The following code ignores the protected mode setting, which is required when we want to ignore the protected mode settings for Internet Explorer:

```
InternetExplorerOptions options = new InternetExplorerOptions();
options.IntroduceInstabilityByIgnoringProtectedModeSettings = true;
InternetExplorerDriver driver = new InternetExplorerDriver(options);
```

autograntpermissions capability on Android

This capability helps Appium to automatically determine the permissions required by a particular app and grants these permissions to the app on install:

```
DesiredCapabilities capabilities = new DesiredCapabilities();
capabilities.setCapability("autoGrantPermissions","true");
```

For iOS, the `autoAcceptAlerts` capability accepts all iOS alerts if they pop up. This includes permission alerts for location, photos, camera, and so on. This works only on the `XCUITest` or `UIAutomation` drivers.

Summary

In this chapter, we covered the topic of window handles, followed by modal and non-modal dialog. We bifurcated pop-up windows and alerts into modal and non-modal dialog and saw examples of these. We explored the Set interface and the simple for loop for handling popups. We created popups and alerts with HTML, CSS, and JavaScript and saw how to handle those using Selenium scripts. Moreover, we had a look at the various methods of the Alert object. We had a look at the `switchTo()` methods and also grasped of handling frames. In this process, we created a sample frame. Finally, we saw what `ChromeOptions` were available and also saw options for Internet Explorer and Firefox.

The next chapter is going to cover one of the most important topics: synchronization.

5
Synchronization

We are going to learn about a very important topic in this chapter: synchronization in Selenium WebDriver. Until now, we have been using `Thread.sleeep` to unconditionally wait for a web element to appear on screen. This is not suitable when network speed is good because the code will keep on waiting until the specified time, even though the element has already been rendered on screen. We will see some examples of the synchronization methods available to us and learn about the differences between each of the methods. We will, as always, dissect each method and study its class hierarchy.

This chapter covers the following topics

- What is synchronization?
- Different types of synchronization
- Creating a sample page in JavaScript
- Understanding implicit wait
- Understanding explicit wait and the `ExpectedConditions` class
- Understanding fluent wait

Technical requirements

You will be required to have Tomcat 6 in addition to the software mentioned in Chapter 4, *Handling Popups, Frames, and Alerts*.

The code files of this chapter can be found on GitHub:
https://github.com/PacktPublishing/Selenium-WebDriver-Quick-Start-Guide/tree/master/Chapter05

Check out the following video to see the code in action:
http://bit.ly/2RjRB9P

What is synchronization?

The dictionary definition of synchronization states that it is a process of coordinating two or more activities at a time. These activities can be anything, such as various instruments playing various tunes in an orchestra or the different parts of an automated assembly line in a car factory.

As far as test automation goes, synchronization is the process of matching the speeds of the application under test and the testing tool, which in this case is Selenium. Synchronization in Selenium WebDriver is implemented using something called a **wait**.

Synchronization types

Let's have a look at the different types of synchronization. There are two types of synchronization:

- **Unconditional synchronization**: In this case, the testing tool waits until the time specified in the wait statement—for example, `Thread.sleep(5000);` is an instruction for Selenium to wait for five seconds (the `sleep` method accepts the time in milliseconds). Say that the web element shows up in 3 seconds. In this case, 2 seconds will be a wasted, as Selenium will keep on waiting for 5 seconds.
- **Conditional synchronization**: In this case, we specify a condition that has to be satisfied before throwing a `NoSuchElementException`. If the specified condition is met, then Selenium will not wait any longer but will move on to execute the next statement. In this chapter, we will look at three types of wait that fall into this category: the implicit wait, explicit wait, and the fluent wait.

Let's look at these different types of synchronization in more detail.

Unconditional synchronization

In most of the examples that we have looked at so far, we have made use of `Thread.sleep(long);` to wait for an element. Someone who is new to Java might wonder what a thread is. Let's first learn about the concept of a thread.

A thread is a single sequence of execution in a program. A particular program always runs on one thread, which is called the **main thread**. Another type of thread is one that is started from the main thread explicitly, and is called the **child thread**. In all the programs that we have seen so far, the execution was done on the main thread.

Thread.sleep(long)

Let's look at the first type of unconditional synchronization, `Thread.sleep`.

The signature of the `sleep(long)` method is `public static native void sleep(long) throws InterruptedException`. There is an overloaded version of the sleep method, `public static void sleep(long, int)`. Usually, the one with a single argument is used.

The `sleep(long)` method is a method in the `Thread` class that makes a particular thread wait for the specified number of milliseconds. When we enter `Thread.sleep(5000);`, we ask the main thread to wait for 5 seconds. Since the main thread stops, execution is halted for 5 seconds, after which the next statement is executed. If you put `Thread.sleep()` in many places in your code, you pause the main thread, which can slow down execution to a considerable extent. Therefore, it is not recommended that you use `Thread.sleep(long)` in your code, and that you instead use conditional synchronization.

Conditional synchronization

Let's jump into conditional synchronization now. There is polling involved in conditional synchronization. There are two categories in conditional synchronization:

- Synchronization at the driver-instance level
- Synchronization at the WebElement level

Let's look at synchronization at the driver-instance level.

Synchronization at the WebDriver-instance level

There are three kinds of wait in this category:

- Implicit wait: Applies to all elements on a web page
- `PageLoadTimeout`: Checks for page load time
- `SetScriptTimeout`: Can be used to set the timeout for an asynchronous script execution

Implicit wait

The implicit wait is the first type of conditional synchronization that is always at the WebDriver-instance level. As long as the WebDriver instance is active, this wait is effective.

The way to use the implicit wait is by phrasing it as `driver.manage().timeouts().implicitlyWait(14, TimeUnit.SECONDS);`.

The `manage()` method, when invoked on the `driver` object, returns an object of the options interface, after which we invoke the `timeouts()` method in the options interface. The `timeouts()` method returns an object of the timeouts interface. Finally, we invoke the `implicitelyWait(long, TimeUnit)` method of the timeouts interface. The `implicitelyWait` method is a factory method that returns an object of the same interface, namely `timeouts`.

When searching for a single element, the driver should poll the page until it finds the element under consideration. If the element is not found, or this timeout expires, a `NoSuchElementException` is thrown. When multiple elements are being searched in the DOM, the driver polls the page until at least one element has been found, or this timeout expires.

Increasing the implicit wait timeout is a method that should be used with caution, as it will have a negative effect on the test runtime, particularly when used with slower location techniques, such as XPath, when compared with CSS.

We will create a sample page in JavaScript, which will have a button. When the button is clicked, various tool names will get displayed at a time interval of 3 seconds. The following is the code for the web page:

```
<html>
 <body>
    <button onclick="timerText()">Timer Start</button>
    <p id="demoText">Click on Timer Start</p>
    <script>
        function timerText() {
            setTimeout(displaySelenium, 3000);
            setTimeout(displayUFT, 6000);
            setTimeout(displayAppium, 9000);
            setTimeout(displaySilkTest, 12000);
            setTimeout(displayCucumber, 15000);
        }
        function displaySelenium() {
            document.getElementById("demoText").innerHTML="Selenium
            WebDriver";
```

```
        }
        function displayUFT() {
            document.getElementById("demoText").innerHTML="Unified
            Functional Test";
        }
        function displayAppium() {
            document.getElementById("demoText").innerHTML="Appium";
        }
        function displaySilkTest() {
            document.getElementById("demoText").innerHTML="Silk Test";
        }
        function displayCucumber() {
            document.getElementById("demoText").innerHTML="Cucumber
            Framework";
        }
    </script>
  </body>
</html>
```

Let's write a small program that prints Element is displayed when it finds an element with the text Silk Test on screen. Note that, as per the preceding JavaScript program, Silk Test appears after 12 seconds.

The following is the code to print details such as whether an Element is displayed and the exact time required to find the element. The Start Time and End Time are noted, and finally, a difference in time is measured:

```
long startTime = 0L;
long endTime = 0L;
boolean status = false;
System.setProperty("webdriver.chrome.driver","C:\\SeleniumWD\\src\\main\\re
sources\\chromedriver.exe");
WebDriver driver = new ChromeDriver();
driver.manage().window().maximize();
driver.navigate().to(
"C:\\Users\\Bhagyashree\\Desktop\\Documents\\Timer.html");
driver.findElement(By.xpath("//*[text()='Timer Start']")).click();
startTime = System.currentTimeMillis();
driver.manage().timeouts().implicitlyWait(20, TimeUnit.SECONDS);
try {
        status = driver.findElement(By.xpath("//*[text()='Silk Test']"))
                .isDisplayed();
    } catch (NoSuchElementException exception) {
        System.out.println(exception.getMessage());
      }
  if (status) {
     System.out.println("Element is displayed");
```

```
    } else {
        System.out.println("Element is not displayed");
    }
endTime = System.currentTimeMillis();
timeDiff = endTime - startTime;
timeDiff = timeDiff / 1000;
System.out.println("Start Time: " + startTime);
System.out.println("End Time: " + endTime);
System.out.println("Difference in Time: " + timeDiff);
```

The output from the preceding code is given here:

```
Element is displayed
Start Time: 1535203696762
End Time: 1535203708848
Difference in Time: 12
```

The page load timeout is a timeout that remains in action for the lifetime of the driver. It sets the amount of time for a page load to complete fully before throwing a `PageLoadTimeout` exception. Let's take the example of http://www.freecrm.com. The home page takes some time to fully load before it can display the login form. As you can see, we get the exact time for `Silk Test` to appear on screen, which is 12 seconds. Note that it is not feasible for us to use `Thread.sleep` in this kind of situation.

The following script displays the use of `PageLoadTimeout`, where we have deliberately set the `PageLoadTimeout` as 3 seconds:

```
System.setProperty("webdriver.chrome.driver",
"C:\\SeleniumWD\\src\\main\\resources\\chromedriver.exe");
WebDriver driver = new ChromeDriver();
driver.manage().timeouts().pageLoadTimeout(3, TimeUnit.SECONDS);
driver.manage().window().maximize();
driver.navigate().to("http://www.freecrm.com");
```

This script will throw a `PageLoadTimeout` exception, as shown in the following output:

```
Driver info: org.openqa.selenium.chrome.ChromeDriver
Capabilities [{platform=XP, acceptSslCerts=false, javascriptEnabled=true, browserName=chrome, chrome={userDataDir=C:\Users\PINAKI~1.CHA\AppData\Local\Temp
    at sun.reflect.NativeConstructorAccessorImpl.newInstance0(Native Method)
    at sun.reflect.NativeConstructorAccessorImpl.newInstance(Unknown Source)
    at sun.reflect.DelegatingConstructorAccessorImpl.newInstance(Unknown Source)
    at java.lang.reflect.Constructor.newInstance(Unknown Source)
    at org.openqa.selenium.remote.ErrorHandler.createThrowable(ErrorHandler.java:206)
    at org.openqa.selenium.remote.ErrorHandler.throwIfResponseFailed(ErrorHandler.java:158)
    at org.openqa.selenium.remote.RemoteWebDriver.execute(RemoteWebDriver.java:595)
    at org.openqa.selenium.remote.RemoteWebDriver.get(RemoteWebDriver.java:306)
    at org.openqa.selenium.remote.RemoteWebDriver$RemoteNavigation.to(RemoteWebDriver.java:850)
    at test.testAutomation.src.helperclasses.PageLoadTimeout.main(PageLoadTimeout.java:16)
```

Another timeout that can be applied at the driver-instance level is `setScriptTimeout`. This timeout requires an understanding of the `JavaScriptExecutor`, which is covered in a later chapter; `setScriptTimeout` will be discussed then.

Synchronization at the WebElement level

We will now learn about synchronization at the `WebElement` level. There are two waits in this category:

- Explicit wait
- Fluent wait

Explicit wait

To understand explicit waits, we have to explore the `WebDriverWait` class. This class extends the `FluentWait` class. The `FluentWait` class implements the wait interface.

The `WebDriverWait` constructor has three overloaded versions. We will be using the version with two arguments, which is the most common version. The signature of the `WebDriverWait` constructor with two arguments is `public WebDriverWait(WebDriver, long)`. The way to instantiate a `WebDriverWait` object is by phrasing the code as follows:

```
WebDriverWait wdWait = new WebDriverWait(driver,15);
```

Here, we specify the maximum waiting time as `15` seconds. To understand the explicit wait, we should first familiarize ourselves with a class called `ExpectedConditions`. This class has a number of useful methods for waiting on a WebElement to appear before throwing an `ElementNotFoundException`. A few of these very useful methods are as follows:

- `presenceOfElementLocated`: Checks whether the element is present on the DOM of a page.
- `visibilityOfElementLocated`: Checks whether the element is present on the DOM of a page and is also visible.
- `frameToBeAvailableAndSwitchToIt`: Checks whether a given frame is available. If it is, it then switches to the frame.
- `elementToBeClickable`: Checks whether a given element is visible and enabled.

The syntax of using these methods is as follows:

```
wdWait.until(ExpectedConditions.presenceOfElementLocated(By.xpath("//*[text
()='Silk Test']")));
```

Let's look at this line of code in a little more detail. First, let's split it. The `until` method is present in the `FluentWait` class. Since the `WebDriverWait` class extends the `FluentWait` class, we can invoke the `until` method on the `WebDriverWait` object. The `until` method keeps polling for the element passed as a parameter until the timeout value specified at the time of creating the `WebDriverWait` object. The default polling time is 500 milliseconds. Selenium will check for the presence of the element every 500 milliseconds before throwing a `NoSuchElement` exception.

Let's have a look at a small program that uses an explicit wait to find particular text when it is visible. We will use the same JavaScript file that was created earlier:

```
long startTime = 0L;
long endTime = 0L;
boolean status = false;
System.setProperty("webdriver.chrome.driver",
"C:\\SeleniumWD\\src\\main\\resources\\chromedriver.exe");
 WebDriver driver = new ChromeDriver();
 WebDriverWait wdWait = new WebDriverWait(driver, 12);
 driver.manage().window().maximize();
 driver.navigate().to(
"C:\\Users\\Bhagyashree\\Desktop\\Documents\\Timer.html");
 wdWait.until(
ExpectedConditions.presenceOfElementLocated(By.xpath("//*[text()='Timer
Start']"))).click();
 startTime = System.currentTimeMillis();
 try {
     wdWait.until(ExpectedConditions.presenceOfElementLocated(By
         .xpath("//*[text()='Silk Test']")));
     status = driver.findElement(By.xpath("//*[text()='Silk Test']"))
.isDisplayed();
 } catch (NoSuchElementException exception) {
     System.out.println(exception.getMessage());
 }
 if (status) {
System.out.println("Element is displayed");
 } else {
System.out.println("Element is not displayed");
 }
 endTime = System.currentTimeMillis();
 timeDiff = endTime - startTime;
 timeDiff = timeDiff / 1000;
 System.out.println("Start Time: " + startTime);
```

```
System.out.println("End Time: " + endTime);
System.out.println("Difference in Time: " + timeDiff);
```

The output from the preceding program is as follows:
Element is displayed
Start Time: 1535263491998
End Time: 1535263504312
Difference in Time: 12

Fluent wait

A fluent wait is the implementation of the wait interface in which we can configure the timeout and the polling interval dynamically. As well as this, it enables us to add exception exclusions, such as the following example, where we ignore the NoSuchElement exception. Each instance of the fluent wait defines the timeout value and the polling time or interval. The polling interval states the frequency with which to check for the presence of the element under consideration.

Let's take the example HTML we created and try to find out whether we are able to locate the element under consideration using fluent wait, using the following code:

```
long startTime = 0L;
long endTime = 0L;
WebElement elem = null;
boolean status = false;
System.setProperty("webdriver.chrome.driver","C:\\SeleniumWD\\src\\main\\
    resources\\chromedriver.exe");
WebDriver driver = new ChromeDriver();
Wait<WebDriver> wdWait = new FluentWait<WebDriver>(driver)
                            .withTimeout(30, TimeUnit.SECONDS)
                            .pollingEvery(1, TimeUnit.SECONDS)
                            .ignoring(NoSuchElementException.class);
driver.manage().window().maximize();
driver.navigate().to("C:\\Users\\Bhagyashree\\Desktop\\Documents\\Timer.htm
l");
 wdWait.until(ExpectedConditions.presenceOfElementLocated(By.xpath(
            "//*[text()='Timer Start']"))).click();
 startTime = System.currentTimeMillis();
 try {
        elem = wdWait.until(new Function<WebDriver, WebElement>() {
            public WebElement apply(WebDriver driver) {
            WebElement text1 = driver.findElement(By
                            .xpath("//*[@id='demoText']"));
            String value = text1.getAttribute("innerHTML");
            if (value.equalsIgnoreCase("Silk Test")) {
                return text1;
```

```
            } else {
                System.out.println("Text on screen: " + value);
                return null;
            }
        }
    });
} catch (NoSuchElementException exception) {
System.out.println(exception.getMessage());
}
if (elem.isDisplayed()) {
    System.out.println("Element is displayed");
    } else {
        System.out.println("Element is not displayed");
    }
endTime = System.currentTimeMillis();
timeDiff = endTime - startTime;
timeDiff = timeDiff / 1000;
System.out.println("Start Time: " + startTime);
System.out.println("End Time: " + endTime);
System.out.println("Difference in Time: " + timeDiff);
```

The output from this program is as follows:
Text on screen: Click on Timer Start
Text on screen: Click on Timer Start
Text on screen: Click on Timer Start
Text on screen: Selenium WebDriver
Text on screen: Selenium WebDriver
Text on screen: Selenium WebDriver
Text on screen: Unified Functional Test
Text on screen: Unified Functional Test
Text on screen: Unified Functional Test
Text on screen: Appium
Text on screen: Appium
Text on screen: Appium
Element is displayed
Start Time: 1535286265688
End Time: 1535286278250
Difference in Time: 12

This output clearly shows that the DOM is polled every second and each piece of text appears three times. The reason for this is that in the JavaScript, the various text items are displayed after a delay of three seconds.

The polling stops when the required text is found and we get the exact time as 12 seconds.

Let's look at two important concepts to bear in mind when using fluent waits.

Function keyword

In the preceding example, we have used `Function` keyword in the form of `new Function<WebDriver, WebElement>()`.

When we use `Function`, we can return any value. In this case, we return a web element. All instances of `Function` should have a single method called `apply`, which in the preceding case takes a `WebDriver` object as a parameter and returns the `WebElement` that is obtained from the `findElement` method. The advantage of using instances of `Function` is that one can pass an object of any type as the input and return an object of any type as the output.

It is not just web elements that can be returned. We can even return a Boolean condition. Boolean conditions of the kind shown in the following code are allowed:

```
return driver.findElements(By.id("xyz")).size() > 0
```

The `apply` method will look like the one shown in the following code when they are used to return a Boolean value:

```
new Function<WebDriver, Boolean>() {
    public Boolean apply(WebDriver driver) {
        //Processing comes here
    }
};
```

Let's have a look at some web pages that are created using JQuery and AJAX. In the following section, we will create a page using the JQuery library.

A glance at the JQuery library

Here, we will introduce a JavaScript library called JQuery. JQuery is one of the most popular JavaScript libraries available today. JQuery makes DOM traversal and manipulation simple. JQuery becomes more powerful when used with **AJAX**. AJAX stands for **Asynchronous JavaScript and XML** and is a web-development technique to send and receive data in the background without refreshing the web page. Many modern websites are being built using a combination of JQuery and AJAX.

JQuery can be used via the **content delivery network** (**CDN**) or downloaded as a JS file. For the example in this chapter, we will be using the CDN approach. We are using the 3.3.1 version of JQuery.

Cross-browser compatibility is a major advantage of JQuery over raw JavaScript. JQuery is extensible and has rich AJAX support.

We will be creating a sample JQuery application for pulling data from a text file on the web server. We will be making use of the JQuery AJAX `load()` function to achieve this.

A Sample application using JQuery

In this section, we will create a very simple JQuery application that will pull data from a text file on the server. We are using Tomcat 6 as the web server. You are free to use a later version, if you wish.

Tomcat setup

Tomcat can be downloaded from `https://tomcat.apache.org/`. Once Tomcat is downloaded, go through the following the steps:

1. Extract the ZIP file to a convenient folder
2. Go to the `bin` directory in the Tomcat installation and double-click `startup.bat`

A console window similar to the one shown in the following screenshot will pop up:

3. When the text `INFO: Server startup in ... ms` is displayed, it means that Tomcat has started successfully. To verify the Tomcat server, simply go to `http://localhost:8080`. A screen similar to the one shown in the following screenshot will then be displayed:

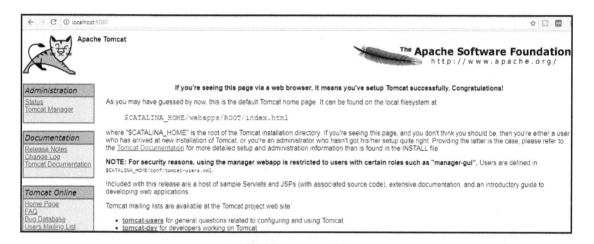

Creating the HTML fileThe next step is to create the HTML and JavaScript files for our sample application.

Next, we create the HTML file with an internal JQuery.

The following code shows the HTML file with the internal JQuery and CSS:

```
<!DOCTYPE html>
<html>
    <head>
    <style>
        #waitlogo {
            border: 16px solid #f3f3f3; /* Light grey */
            border-top: 16px solid #3498db; /* Blue */
            border-radius: 50%;
            width: 120px;
            height: 120px;
            animation: spin 2s linear infinite;
        }
        @keyframes spin {
            0% { transform: rotate(0deg); }
            100% { transform: rotate(360deg); }
        }
    </style>
```

```
<script
src="https://ajax.googleapis.com/ajax/libs/jquery/3.3.1/jquery.min.js"></sc
ript>
<script>
    $(document).ready(function(){
    $(document).ajaxStart(function(){
        $("#waitlogo").css("display", "block");
    });
     $(document).ajaxComplete(function(){
     $("#waitlogo").css("display", "none");
     });
     $("button").click(function(){
        $("#txt").load("demo_ajax_load.txt");
     });
     });
</script>
</head>
<body>
    <div id="txt"><h2>Ajax will change the page contents</h2></div>
    <button id="changecontent">Change Page Content</button>
    <div id="waitlogo"
style="display:none;width:69px;height:89px;border:1px          solid
black;position:absolute;top:50%;left:50%;padding:2px;"><img
src='giphy.gif' width="64" height="64" /><br>Loading..</div>
    </body>
</html>
```

(`giphy.gif` has been taken from `https://giphy.com/`)

We do not need to go into the details of the preceding snippet. This HTML should be placed in the `webapps/ROOT` folder of the Tomcat installation. The page that is generated is shown in the following screenshot:

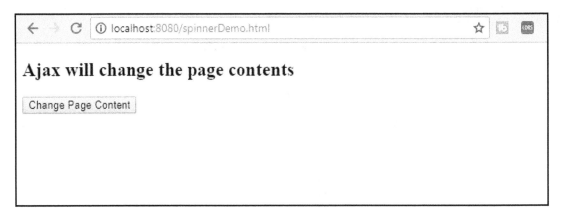

Upon clicking the **Change Page Content** button, a spinner will load while data is fetched from `demo_ajax_load.txt`, which is also present in the same Tomcat directory as the `html` file, as shown in the following screenshot:

Once the AJAX call completes, the data from the text file is displayed on screen.

The Selenium script to check whether any active JQuery request is currently going on is shown in the following code:

```
System.setProperty("webdriver.chrome.driver",
"C:\\SeleniumWD\\src\\main\\resources\\chromedriver.exe");
RemoteWebDriver driver = new ChromeDriver();
driver.navigate().to("http://localhost:8080/spinnerDemo.html");
driver.findElement(By.xpath("//*[@id='changecontent']")).click();
Boolean isJqueryCallDone = false;
int counter = 0;
while (counter < 10) {
    isJqueryCallDone = (Boolean) driver
        .executeScript("return jQuery.active==0");
    System.out.println("JQuery call in loop: " + isJqueryCallDone);
    if (isJqueryCallDone == true) {
        break;
    } else {
        counter++;
    }
}
System.out.println("JQuery call done: " + isJqueryCallDone);
```

The output is shown in the following printout:
JQuery call in loop: false

```
JQuery call in loop: false
JQuery call in loop: false
JQuery call done: true
```

The output prints `false` in the console until the JQuery has completed the AJAX call. The duration of the AJAX call depends on the size of the `demo_ajax_load.txt`. During this execution, the size of the text file was 4,692 KB.

Exposing the JavascriptExecutor

As we can see in the preceding code, there is a new entry in our encyclopedia of Selenium concepts, which is `JavascriptExecutor`. It is one of the interfaces in Selenium that can be used to accomplish things that cannot be accomplished through regular locators. It lets us execute JavaScript code in Selenium. Let's study the following line from the code in the previous section:

```
isJqueryCallDone = (Boolean) ((JavascriptExecutor)
driver).executeScript("return jQuery.active===0");
```

The `isJqueryCallDone` phrase is a Boolean variable that is assigned the value of `return jQuery.active===0`.

Here, we are invoking the `executeScript` function on the `driver` object after casting it to `JavascriptExecutor`. Only then will the driver able to execute the JavaScript commands.

We will see more of `JavascriptExecutor` in a later chapter, where we will do a thorough examination of `JavascriptExecutor`.

Pitfall – Never fall into one

Sometimes, you might feel that to be on the safe side, adding a little implicit wait in addition to the explicit wait will help. Never do this! This can have unpredictable results. Doing this may increase waiting time, or may give you an `ElementNotFoundException`.

Decide on a particular wait and use that throughout; this is the best strategy to apply.

Using a combination of implicit and explicit waits will make your explicit waits break and produce unpredictable results.

Summary

In this chapter, we had a look at the various kinds of synchronization. We started off with `Thread.sleep` and learned how it is not feasible in most situations. We looked at explicit waits and the `ExpectedConditions` class. We also learned about fluent waits, and learned how to configure the polling time and use it to display text at intervals. We explored the `Function` and 'predicates' components of fluent waits.

Finally, we established a best practice of not using implicit and explicit waits together. Doing so can have an unpredictable impact on the execution time of your script, and can even break explicit waits.

In the next chapter, we will dive deep into the `Actions` class and `JavascriptExecutor` in detail.

6
The Actions Class and JavascriptExecutor

All of the concepts we have studied so far work well for simple operations such as entering values in a text box, selecting values from a drop-down, and clicking radio buttons and checkboxes, but a time comes when we need to perform advanced interactions on the screen such as double-click, drag and drop, and right-click. This is when the `actions` class and `JavascriptExecutor` come to the rescue. Throughout this chapter, we will see examples of the `actions` class and `JavascriptExecutor`.

In this chapter, we will cover the following topics:

- Introducing the builder pattern
- Introducing the `actions` class
- Various examples of the `actions` class
- Introducing `JavascriptExecutor`
- Various examples of `JavascriptExecutor`
- Scratching the surface of `EventFiringWebDriver`
- Getting started with the `BaseClass` framework
- Introducing the Selenium Grid

Let's begin by learning about the builder pattern.

Technical requirements

You will be required to have Selenium-Server-Standalone 3.14 in addition to the software in Chapter 5, *Synchronization*.

The code files of this chapter can be found on GitHub:
https://github.com/PacktPublishing/Selenium-WebDriver-Quick-Start-Guide/tree/master/Chapter06

Check out the following video to see the code in action:
http://bit.ly/2D7sIet

Builder design pattern

The builder pattern is a creational design pattern in the huge collection of design patterns. It is used to create complex objects from a collection of other objects. This complex object is built in a step-by-step manner and in such a way that the constituent objects are unaware of what the final object is.

A car is an ideal example of the builder pattern. The functionality of any car is almost the same. It has many components working together, such as the engine, radiator, and brake system. The radiator, as a standalone object, does not know what the final object is going to be. It may be any kind of car. The radiator can be considered a comparatively simple object when compared with the final product that is any car.

Another example of the builder pattern is a laptop. The functionality of the laptop is quite complex, but the constituent parts, such as the keyboard, are comparatively simple.

We do not have to go into the details of the builder pattern here, since a basic understanding is sufficient. We will go into details as and when required.

The following are the advantages of the builder pattern:

- Distinction between the construction and the final representation of the object
- The constituent parts can be changed
- Better hold on the construction process

Let's see an implementation of this design pattern in the actions class.

The actions class

Websites, nowadays, have various features such as double click, drag and drop, and hover, to accomplish certain functionality. A special API called the Actions API is used to handle these scenarios. The `actions` class implements the builder pattern by creating a composite action made up of constituent actions that are specified by method calls.

Actions is a class in `org.openqa.selenium.interactions.Actions` and it extends the base class object. This class has one protected field called **action**, which is of the `CompositeAction` type. `CompositeAction` is a class that collects all actions and triggers them at the same time. `CompositeAction` implements the Action interface. The `action` interface has just one method: `perform()`, which the `actions` class implements.

There are various methods in the `actions` class. All of these methods return an actions object unless specified explicitly. The generic code for a few of the important methods in this class are shown when we look at the `TestBase` class, and are as follows:

- `keyDown(Keys)`: Simulates a key press. The key under consideration remains pressed and the subsequent operations may assume that the key is kept pressed.
- `keyDown(WebElement,Keys)`: Simulates a key press after focusing on a `WebElement`. The key under consideration remains pressed and the subsequent operations may assume that the key is kept pressed.
- `keyUp(Keys)`: Simulates a key release. If you try to use this method on an undepressed key, this will yield an undefined behavior.
- `keyUp(WebElement, Keys)`: Simulates a key release after focusing on an element.
- `sendKeys(CharSequence...)`: Sends a variable number of arguments of the `CharSequence` type. This is a key combination that's sent to the currently active element.
- `sendKeys(WebElement, CharSequence...)`: Sends a variable number of arguments of the `CharSequence` type after focusing on the `WebElement` that's passed as a parameter.
- `clickAndHold()`: Simulates a left mouse click without releasing at the current mouse location.
- `clickAndHold(WebElement)`: Simulates a left mouse click without releasing in the center of a given web element.

- `release()`: Releases the depressed left mouse button at the current location. Attempting to invoke this method without first invoking `clickAndHold()` will result in an undefined behavior.
- `release(WebElement)`: Releases the depressed left mouse button in the middle of the web element. Attempting to invoke this method without first invoking `clickAndHold()` will result in an undefined behavior.
- `click()`: Left mouse click at the current mouse location.
- `click(WebElement)`: Left mouse click in the middle of the given web element.
- `doubleClick()`: Performs a double click at the current mouse location.
- `doubleClick(WebElement)`: Performs a double click at the middle of the given web element.
- `moveToElement(WebElement)`: Moves the mouse cursor to the center of the element. The element is also scrolled into view.
- `moveToElement(WebElement,int,int)`: Moves the mouse cursor to an offset position from the top-left corner of the element. The element is also scrolled into view.
- `moveByOffset(int,int)`: Moves the mouse from its current position (or 0,0 if the mouse has not been moved) by the given offset. If the coordinates that are provided are outside the visible window (the mouse will end up outside the browser visible window), then the content is scrolled to match.
- `contextClick()`: Performs a right-click at the current mouse location.
- `contextClick(WebElement)`: Performs a right-click at the center of the given element. First, it does a `mouseMove` to the location of the element.
- `dragAndDrop(WebElement, WebElement)`: Executes click-and-hold at the source element, moves to the location of the target element, and then releases the mouse.
- `dragAndDropBy(WebElement,int,int)`: Executes click-and-hold at the source element, moves the mouse by a given offset, and then releases the mouse.
- `build()`: Generates a composite action containing all of the actions that have been gathered so far, ready to be performed (the internal builder state is reset, so subsequent calls to `build()` will contain fresh sequences).
- `perform()`: Performs the accumulated actions without calling `build()` first. This method has a return type void.

Various scenarios for the actions class

In this section, we will take a look at two scenarios for the `actions` class: double click and hover.

For the double click scenario, the following code creates an HTML page. This HTML page uses JQuery to show and hide the contents of a paragraph:

```
<!DOCTYPE html>
<html>
<head>
 <title>JQuery Demo</title>
 <script
     src="https://code.jquery.com/jquery-3.3.1.js"
     integrity="sha256-2Kok7MbOyxpgUVvAk/HJ2jigOSYS2auK4Pfzbm7uH60="
     crossorigin="anonymous"></script>
 <script type="text/javascript">
     $(document).ready(function(){
         $('h2').dblclick(function(){
             $('p').show();
         })
     });
 </script>
 <style>
    p {
         display:none;
    }
 </style>
</head>
<body>
 <h2>Double click Here!</h2>
 <p>This is some sample text</p>
</body>
</html>
```

Next, we put in a Java class with Selenium code. The following code uses the `actions` class instance to double click the h2 tag. On double clicking, the paragraph tag is displayed and the text in the paragraph tag is captured:

```
public class JQueryDBLClick {
 public static void main(String[] args) {
 String userDirectory = System.getProperty("user.dir");
 System.setProperty("webdriver.chrome.driver",
 userDirectory+"\\src\\main\\resources\\chromedriver.exe");
 WebDriver driver = new ChromeDriver();
 driver.manage().timeouts().implicitlyWait(15, TimeUnit.SECONDS);
 driver.navigate().to(
```

```
"C:\\Users\\Bhagyashree\\Documents\\DoubleClick.html");
Actions actions = new Actions(driver);
actions.doubleClick(driver.findElement(By.tagName("h2"))).perform();
System.out.println("Text in para: "
+ driver.findElement(By.tagName("p")).getText());
  }
}
```

 You can also set the path of the binary executable in the `path` variable. Hence, you can avoid using `System.setProperty()` to set the path explicitly.

Next, we will explore the hover scenario, and for this, we will make use of https://jqueryui.com/tooltip/, which has a ready-made tooltip that we can make use of. It can be seen that the `aria-describedby` HTML attribute is only inserted into the HTML when a mouse hover is done on this element.

This code extracts the `aria-describedby` attribute and prints it to the console:

```
public class HoverDemo {
    public static void main(String[] args) {
        String userDirectory = System.getProperty("user.dir");
        System.setProperty("webdriver.chrome.driver",
            userDirectory+"\\src\\main\\resources\\chromedriver.exe");
        WebDriver driver = new ChromeDriver();
        driver.manage().timeouts().implicitlyWait(15,
        TimeUnit.SECONDS);
        driver.navigate().to("https://jqueryui.com/tooltip/");
        driver.manage().window().maximize();
        Actions actions = new Actions(driver);
        driver.switchTo().frame(0);
        actions.moveToElement(driver.findElement(By.id("age"))).build()
            .perform();
        System.out.println("Attribute is: "
            + driver.findElement(By.id("age")).getAttribute(
                "aria-describedby"));
  }
}
```

Introducing JavascriptExecutor

`JavascriptExecutor` is is an interface in Selenium, which indicates that a WebDriver instance can execute raw JavaScript, providing access to the mechanism for doing that.

It has two methods:

- The `executeScript(String,Object...)` object: Executes JavaScript code in the context of the currently selected frame or window. The script fragment provided will be executed as an anonymous function.
- The `executeAsyncScript(String, Object...)` object: Executes an asynchronous JavaScript in the context of the currently selected frame or window. In contrast to executing synchronous JavaScript, scripts executed with this method must specifically send a signal when they are complete by invoking the associated callback function. This callback function is injected into the executed function as the last argument.

Various scenarios for JavascriptExecutor

Let's have a look at few of the most common scenarios for `JavascriptExecutor`. All of the reusable methods are included in `TestBase.java`:

- The following code waits for the page or document to load completely:

```java
public static boolean docState() {
docReady = ((JavascriptExecutor) driver)
  .executeScript("return document.readyState").toString()
.equals("complete");
return docReady;
}
```

- The following code waits for an AJAX call to complete:

```java
public static boolean ajaxState() {
    int counter = 0;
    while (counter < 10) {
      ajaxIsComplete = (Boolean) ((JavascriptExecutor) driver)
          .executeScript("return jQuery.active == 0");
      if (ajaxIsComplete) {
        break;
      } else {
        counter = counter + 1;
      }
    }
}
```

```
            return ajaxIsComplete;
    }
```

- The following code sets the focus and scrolls into view:

```
public static void executeFocus(String elementID) {
    jse = (JavascriptExecutor) driver;
    jse.executeScript("document.getElementById('" + elementID
        + "').focus();");
}
```

```
public static void scrollintoviewJS(WebElement element) {
    jsscript1 = "arguments[0].scrollIntoView(false);";
    js.executeScript(jsscript1, element);
}
```

- The following code sets the style visibility by `id`:

```
public static void setStyleVisibility(String id) {
    jse = (JavascriptExecutor) driver;
    jse.executeScript("document.getElementById('" + id
        + "').style.visibility = 'visible';");

}
```

- The following code sets the style block by `id` and `name`:

```
public static void setStyleVisibilityByidBlock(String id) {
    jse = (JavascriptExecutor) driver;
    jse.executeScript("document.getElementById('" + id
                + "').style.display='block';");
}
public static void setStyleVisibilityBynameBlock(String name) {
    jse = (JavascriptExecutor) driver;
    jse.executeScript("document.getElementsByName('"+name+"')
                [0].style.display='block';");
}
```

Let's see the scroll in action with an example.

The following code opens the JQueryUI URL, scrolls down to the `Easing` hyperlink, and clicks it:

```
public class ScrollintoView {
    public static void main(String[] args) {
        String userDirectory = System.getProperty("user.dir");
        System.setProperty("webdriver.chrome.driver",
            userDirectory+"\\src\\main\\resources\\chromedriver.exe");
```

```
WebDriver driver = new ChromeDriver();
driver.manage().timeouts().implicitlyWait(15,
TimeUnit.SECONDS);
driver.navigate().to("https://jqueryui.com");
driver.manage().window().maximize();
JavascriptExecutor js = (JavascriptExecutor) driver;
WebElement txtEasing = driver.findElement(By
    .xpath("//a[text()='Easing']"));
js.executeScript("arguments[0].scrollIntoView(true);",
txtEasing);
txtEasing.click();
    }
}
```

The following piece of code will use the same HTML (DoubleClick.html) and, without double clicking the h2 header, it will display the paragraph by altering the CSS appropriately:

```
public class GetHiddenText {
    public static void main(String[] args) {
        String userDirectory = System.getProperty("user.dir");
        System.setProperty("webdriver.chrome.driver",
            userDirectory+"\\src\\main\\resources\\chromedriver.exe");
        RemoteWebDriver driver = new ChromeDriver();
        driver.manage().timeouts().implicitlyWait(15,
        TimeUnit.SECONDS);
        driver.navigate().to(
            "C:\\Users\\Bhagyashree\\Documents\\DoubleClick.html");
        driver.manage().window().maximize();
        String txtPara = driver.findElement(By.id("txtPara"))
            .getAttribute("id");
        driver.executeScript("document.getElementById('" + txtPara
            + "').style.display='block';");
    }
}
```

Next, we will move on to EventFiringWebDriver.

EventFiringWebDriver

For understanding `EventFiringWebDriver`, we need to understand the concept of listeners.

Listeners work silently without hampering the flow of the Selenium WebDriver script. Listeners help us find out the result of a particular action before or after the action happens. Take the scenario of clicking on a button. Listeners indicate when the WebDriver instance is about to execute the click of a button. Similarly, listeners also indicates that a button has already been clicked.

To create such a listener, we need to use an interface called `WebDriverEventListener`, which has to be implemented by a concrete class, which becomes our listener. Let's see the process of creating one.

The following class implements `WebDriverEventListener`. Only the methods used in this example are mentioned here:

```
public class EventHandler implements WebDriverEventListener {
 public void afterChangeValueOf(WebElement elem, WebDriver driver) {
 System.out.println("afterChangeValueOf has been invoked for "
 + elem.toString());
 }

 public void afterClickOn(WebElement elem, WebDriver driver) {
 System.out.println("afterClickOn has been invoked for " +
elem.toString());
 }

 public void afterFindBy(By elem, WebElement arg1, WebDriver driver) {
 System.out.println("FindBy has been triggered for " + elem.toString());
 }

 public void afterNavigateBack(WebDriver driver) {
 System.out.println("afterNavigateBack has been triggered to go back to "
 + driver.getCurrentUrl());
 }

 public void afterNavigateForward(WebDriver driver) {
 System.out.println("afterNavigateForward has been triggered to go to "
 + driver.getCurrentUrl());
 }

 public void afterNavigateTo(String str1, WebDriver driver) {
 System.out.println("afterNavigateTo has been triggered for " + str1);
 }
```

```java
public void afterScript(String str1, WebDriver driver) {
System.out.println("afterScript has been triggered for " + str1);
}

public void beforeChangeValueOf(WebElement elem, WebDriver driver) {
System.out.println("beforeChangeValueOf has been triggered for web
element");
}

public void beforeClickOn(WebElement elem, WebDriver driver) {
System.out.println("Going to click on " + elem.toString());
}

public void beforeNavigateBack(WebDriver driver) {
System.out.println("beforeNavigateBack has been triggered to go back to "
+ driver.getCurrentUrl());
}

public void beforeNavigateForward(WebDriver driver) {
System.out.println("About to trigger beforeNavigateForward "
+ driver.getCurrentUrl());
}

public void beforeNavigateTo(String str1, WebDriver driver) {
System.out.println("About to trigger beforeNavigateTo " + str1);
}

public void beforeScript(String str1, WebDriver driver) {
System.out.println("About to trigger beforeScript " + str1);
}

public void onException(Throwable throwable1, WebDriver driver) {
System.out.println("Exception " + throwable1.getMessage() + " has
occurred");
}

public void beforeFindBy(By, WebElement elem, WebDriver driver) {
System.out.println("About to find element " + by.toString());
}
}
```

1. The following code creates a regular WebDriver:

```java
RemoteWebDriver driver = new ChromeDriver();
```

2. The following code creates an `EventFiringWebDriver` using the driver instance from *Step 1*:

```
EventFiringWebDriver eventDriver = new
EventFiringWebDriver(driver);
```

3. The following code creates an instance of your `eventHandler` class and registers it for events using the register method of the `EventFiringWebDriver` instance:

```
EventFiringWebDriver object created above
EventHandler handler = new EventHandler();
eventDriver.register(handler);
```

4. The following code creates a sample test for `EventFiringWebDriver`:

```
public class EventHandlerTest {
    public static void main(String[] args) throws IOException {
   System.setProperty(Utilities.getProp(Global_Constants.DRIVER,
Global_Constants.CONFIGPATH),
Utilities.getProp(Global_Constants.DRIVERPATH,
Global_Constants.CONFIGPATH));
Map<String, String> capabilities = new HashMap<String, String>();
   capabilities.put(Global_Constants.pageLoadStrategy_key,
                     Global_Constants.pageLoadStrategy_val);
        ChromeOptions options = new ChromeOptions();
        //Disable the Password Confirmation box
        Map<String, Object> prefs = new HashMap<String, Object>();
        prefs.put("credentials_enable_service", false);
        prefs.put("profile.password_manager_enabled", false);
        options.addArguments("disable-extensions");
        options.addArguments("--start-maximized");
        options.setExperimentalOption("prefs", prefs);
        WebDriver driver = new ChromeDriver(options);
        EventFiringWebDriver eventDriver = new
        EventFiringWebDriver(driver);
        EventHandler handler = new EventHandler();
        eventDriver.register(handler);
        eventDriver.get("https://www.google.com");
        WebElement element = eventDriver.findElement(By.id("lst-
        ib"));
        element.sendKeys("Selenium");
    }
}
```

The output from this code is given here:

About to trigger beforeNavigateTo https://www.google.com

afterNavigateTo has been triggered for https://www.google.com

About to find element lst-ib

FindBy has been triggered for lst-ib

We will define a few global constants next. These will be defined in an interface called `Global_Constants`:

```
public interface Global_Constants {
  public static final String CONFIGPATH = System.getProperty("user.dir")
  + "\\src\\test\\testAutomation\\config\\config.properties";
  public static final String ORPATH = System.getProperty("user.dir")
  + "\\src\\test\\testAutomation\\Object_Repository\\OR.properties";
  public static final String DRIVER = "DRIVER";
  public static final String BROWSER = "BROWSER";
  public static final String DRIVERPATH = "DRIVERPATH";
  public static final String pageLoadStrategy_key = "pageLoadStrategy";
  public static final String pageLoadStrategy_val = "eager";
}
```

Next, we define our `config.properties` file. This will be a collection of key-value pairs:

```
DRIVER=webdriver.chrome.driver
DRIVERPATH=/src/test/testAutomation/resources/chromedriver.exe
FRAMEWORKFILEPATH=/src/test/testAutomation/resources/Framework.xls
EXECUTIONRESULTSPATH=/src/test/testAutomation/results/Execution_Result
s.csv
EXTENTREPORTPATH=/src/test/testAutomation/results/Extent_Report.html
SCREENSHOTPATH=/src/test/testAutomation/images/
CHROMESERVERPATH=/Resources/chromedriver.exe
UPDATEFILEPATH=/Resources/update.xls
CHECKFLAGPATH=/Resources/CHECK_FLAG.txt
CLOSEFLAGPATH=/Resources/CLOSE_FLAG.txt
LOGPROPERTIESPATH=/log4j.properties
USERACTIONPATH=/Resources/USER_ACTION_SUMMARY.txt
BROWSER=chrome
```

First steps toward the framework

Create the following packages in `src/main/java`:

- `org.packt.invokers`
- `org.packt.receivers`
- `org.packt.command`
- `org.packt.client`
- `org.packt.testbase`

In the `org.packt.testbase` package, create our `TestBase` class. This will serve as the base class for all of our classes.

This base class has methods for the following:

- Sendkeys
- Sendkeys using `JavascriptExecutor`
- Double click
- Context click

You can add additional methods to this test base and the changes will percolate throughout the framework since all of our classes will extend the test base. The following code shows the declarations, the various methods for sending different keystrokes, and clicking elements:

```
public class TestBase {
  public static WebDriver driver = null;
  protected static Logger log = null;
  protected static Utilities util = null;
  protected static Actions actions = null;
  protected static JavascriptExecutor js = null;
  protected static WebDriverWait wdWait = null;
  static String reportFileName = null;
  protected static Map<String, String> resultMap = new HashMap<String,
  String>();
  protected static Fillo fillo = null;
  protected static Recordset recordSet = null;
  protected static Connection connection = null;
  protected static ExtentReports extentReport;
  protected static ExtentTest extentTest;
  private static Object tempElement;
  private static String firstXPath = "//*[text()='";
  private static String lastXPath = "']";
```

```
 public TestBase() throws IOException, FilloException {
 }

 public static void sendKeys(WebElement element, String text) {
 try {
 wdWait.until(ExpectedConditions.visibilityOf(element));
 element.clear();
 element.sendKeys(text);
 } catch (NoSuchElementException e){
 throw new NoSuchElementException();
 }
 }


 public static void sendKeysTAB(WebElement element, Keys text) {
    element.sendKeys(text);
 }

 public static void clickElement(WebElement element) {
    wdWait.until(ExpectedConditions.elementToBeClickable(element)).click();
 }
```

The following code shows the various methods for selecting data from drop-downs. Simply append this code to the preceding code:

```
public static void selectDrpdwnData(WebElement element, String drpdwnData)
{
    new Select(element).selectByVisibleText(drpdwnData);
}

 public static void selectDrpdwnDataByText(WebElement element,
 String drpdwnData) {
 List<WebElement> optionList = new Select(element).getOptions();
 for (WebElement options : optionList) {
 if (options.getText().equalsIgnoreCase(drpdwnData)) {
 options.click();
 }
    }
 }

 public static void selectDrpdwnDataByText2(WebElement element,
    String drpdwnData) {
   List<WebElement> optionList = new Select(element).getOptions();
   for (WebElement options : optionList) {
     if (options.getText().equalsIgnoreCase(drpdwnData)) {
       options.click();
     }
   }
```

```
    }

    public static void selectDrpdwnDataByCTRL(WebElement element,
        String drpdwnData) {
        List<WebElement> optionList = new Select(element).getOptions();
        actions = new Actions(driver);
        actions.keyDown(Keys.CONTROL);
        for (WebElement options : optionList) {
        options.click();
      }
      actions.keyUp(Keys.CONTROL);
    }
```

The following code is used for double click, right-click, and moving the mouse to a particular element:

```
public static void doubleClickElement(WebDriver driver, WebElement element)
        throws InterruptedException {
        scrollintoviewJS(element);
        System.out.println(element.getText());
        Actions action1 = new Actions(driver).doubleClick(element);
        action1.perform();
    }

    public static void rightClickElement(WebDriver driver, WebElement
    element) {
        actions = new Actions(driver);
        actions.contextClick(element).perform();
    }

    public static void moveToElement(WebDriver driver, WebElement
     element) {
        actions = new Actions(driver);
        actions.moveToElement(element)
        actions.perform();
    }
```

The following code is used to set a text box value, enable the textbox, enter a textbox value, click an element, scroll into the view, and create a web element from its text:

```
public static void setValueJS(String element, String text) {
    String jsscript1 = "document.getElementsByName('" + element
 + "')[0].value='" + text + "';";
    js.executeScript(jsscript1);
  }
```

```
public static void enableTextBoxJS(String element) {
    String jsscript1 = "document.getElementsByClassName('" + element
        + "')[0].enabled='true';";
    System.out.println(jsscript1);
    js.executeScript(jsscript1);
}

public static void enterTextBoxValueJS(String element, String value) {
    String jsscript1 = "document.getElementsByClassName('" + element
        + "')[0].value='" + value + "';";
    System.out.println(jsscript1);
    driver.executeScript(jsscript1);
}

public static void clickElementJS(WebElement element) {
    String jsscript1 = "arguments[0].click();";
    js.executeScript(jsscript1, element);
}

public static void scrollintoviewJS(WebElement element) {
    String jsscript1 = "arguments[0].scrollIntoView(false);";
    js.executeScript(jsscript1, element);
}

public static List<WebElement> findWebElementList(String textFill)
        throws NoSuchElementException {
    tempElement = driver.findElements(By.xpath(firstXPath + textFill
 + lastXPath));
    return (List<WebElement>) tempElement;
}
}
```

These are just the methods that are required for now. We will add more methods as and when required.

If the preceding code does not work and fails on `ChromeDriver` initiation, try updating your version of Guava in the `pom.xml` file to 22.

Understanding the Selenium Grid

Up until now, we have been executing Selenium code on our local machine. But that is not a very good idea. Eventually, we are going to run our framework from Jenkins and moreover, we will be running multiple instances of Chrome, Firefox, and Internet Explorer. Here's where Selenium Grid comes to our rescue.

Selenium Grid can be used to execute code on another machine or a virtual machine. It can also be used to execute code on multiple machines. Code is executed on multiple nodes due to which the total execution time is reduced.

Let's take a look at the Selenium Grid architecture.

Architecture of the Selenium Grid

To understand Selenium Grid, we need to understand the concept of hub and nodes. There is only one hub, but there can be multiple nodes.

The hub is the central machine which sends commands to the respective nodes based on parameters.

The node is the machine where our test will be executed.

The following diagram shows the architecture of Selenium Grid. This diagram shows that we have one hub and there are four nodes: one for Mac, one for Windows, one for Linux, and one for Android:

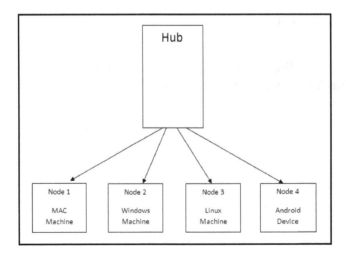

Let's move on to the basic setup of Selenium Grid.

Basic setup

In order to use Selenium Grid, we require a `.jar` file, which has to be downloaded. The file is `selenium-server-standalone-3.14.0.jar`, and is available from `https://www.seleniumhq.org/download/`.

Follow the instructions below for setting up a hub

- Place the `.jar` file in a convenient location
- Navigate to that location in Command Prompt and type in the `java -jar selenium-server-standalone-3.14.0.jar -role hub` command

By giving this command, we are telling the standalone server to start a hub. The command prompt should display the following output:

In the browser, navigate to `http://localhost:4444`.

If you see the following screen, then everything is set correctly and you are ready to configure nodes to the hub:

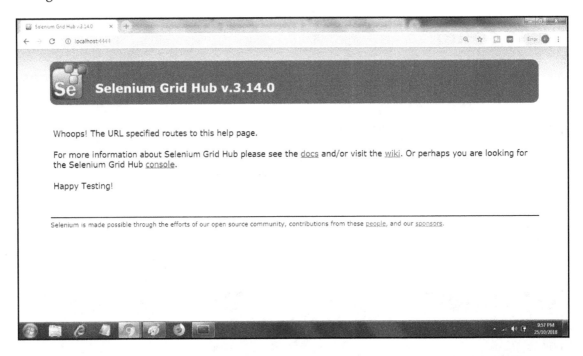

This completes Selenium Grid's basic setup. We will configure nodes in a subsequent chapter.

This completes a very important chapter from a framework standpoint. We have laid the foundation for our framework in this chapter.

Exercise

For the `EventHandler`, try filling the body of the other methods that have not been shown here.

Summary

This chapter started off with the builder pattern, and we went over the `actions` class in detail. We created some generic methods that were put into `TestBase` at the end of this chapter. `TestBase` serves as a parent for all classes in our framework. This is where inheritance comes into the picture. We saw what `JavascriptExecutor` is and also created a few generic methods using this API. We went over a few scenarios of the `actions` class and `JavaScriptExecutor`. We saw the very important concept of listeners in Selenium and went over `EventFiringWebDriver` to listen to the executing code and log it to the console. We also downloaded the Selenium Server standalone for the setup of Selenium Grid and went over its architecture.

In the next chapter, we will look at the command pattern and the various pieces of the framework.

7
The Command Pattern and Creating Components

Learning about Selenium without the knowledge of designing frameworks is not of much use. Frameworks help us with grouping components together and wiring them as per our needs. With a proper and structured framework, adding test cases is just a matter of updating the Excel sheet that has all of the test cases and test steps. It is just a matter of keeping track of the keys in the framework, which in our case is the test case ID.

In this chapter, we will cover the following topics:

- Understanding the command design pattern
- The project structure in Eclipse
- Introducing the TestNG framework and `TestNG.xml`
- Configuring nodes on the Selenium Grid for remote execution

The framework that we will be designing will be based on the command pattern. Hence, it will be a good idea to understand the command pattern first.

Technical requirements

You will be required to have software from `Chapter 6`, *The Actions Class and JavascriptExecutor*.

The code files of this chapter can be found on GitHub:
`https://github.com/PacktPublishing/Selenium-WebDriver-Quick-Start-Guide/tree/master/Chapter07`

Check out the following video to see the code in action:
`http://bit.ly/2EMLKbt`

Introducing the command design pattern

The command design pattern is a behavioral design pattern that involves four players, as in the following example:

- Client
- Invoker
- Command
- Receiver

In very simple words, the receiver is the object on screen such as a textbox and button and the command is `sendKeys` and `click`. We encapsulate the command in an object and pass it to the receiver.

The actual command design pattern also supports an `Undo` operation, but we will not consider the `Undo` operation here. By using command objects, it becomes easier to build generic components that need to assign, delegate, or execute method calls without needing to know the class of the method or the method parameters. Using an `invoker` object allows command executions to be conveniently performed. Using this, we can also create different modes for commands that are handled by the `invoker` object without the client needing to be aware of the existence of the different modes involved.

The main players in a command design pattern are the Client, the invoker, the command, and the receiver. Let's understands all four from a framework perspective.

Client

The client will do the following things:

- Declare references of two abstract classes, such as `AActionKeyword` and `ACommand`. We will take a look at both of these abstract classes later in this chapter.
- Create `invoker` objects.
- Invoke the different commands such as `open`, `navigate`, and `enterText` using the `invoker` objects.

The client will initially call ReadExcelData by passing it the browser name on which the execution has to be done. ReadExcelData will extract records from the Excel sheet, put those in a list of Hashmaps, and pass it back to the client. The client now reads the list of HashMaps, pulls out one Hashmap at a time, extracts the contents, and based on the action keyword executes the switch case logic, which we will look at later.

Invoker

There will be as many invokers as there are are components. Components can be textboxes, radio buttons, and so on. Each individual invoker will create a command object inside it. There will be customized methods inside each invoker, but inside each customized method, the execute method from the ACommand abstract class will be called. The individual invoker will call the overloaded execute methods from the ACommand abstract class.

The following code shows the BrowserInvoker class:

```
public class BrowserInvoker {
  private ACommand aCommand = null;

  public BrowserInvoker(ACommand command) {
    this.aCommand = command;
  }

  public WebDriver open(List<String> browserName) {
    WebDriver driver = aCommand.execute(browserName);
    return driver;
  }
}
```

Command

The command, in our case, will be an abstract class that will have overloaded versions of the execute method. This has been done so that, if we want to create a concrete command class, we don't need to add code to all of the methods of the ACommand interface. The concrete command classes will have the AActionKeyword object, and there will be execute methods that are required for that particular operation. For example, if the operation is a click operation, then there will be just the usage of the no argument execute method.

Next comes the dummy class, ACommand, which has empty implementations.

The following code shows the ACommand abstract class:

```java
public abstract class ACommand {
  public void execute(String x) {
  }
  public void execute(WebDriver driver, String x) {
  }
  public void execute(WebElement element, String x) {
  }
  public WebDriver execute(List<String> x) {
    return null;
  }
  public WebDriver execute() {
    // TODO Auto-generated method stub
    return null;
  }
}
```

Finally, the following code shows the concrete class which creates a new browser session. In the next chapter, the no-argument execute method will be replaced by a single argument method, which includes the browser name:

```java
public class NewBrowser extends ACommand {
  private AActionKeyword actionKeyword = null;
  static WebDriver driver;

  public NewBrowser(AActionKeyword actKeyword) {
    this.actionKeyword = actKeyword;
  }

  public WebDriver execute(List<String> browsers) {
    driver = actionKeyword.openBrowser(browsers);
    return driver;
  }

}
```

Receiver

The Receiver will be an `abstract` class called `AActionKeyword` that will contain empty method bodies that will be placeholders to perform the work by invoking Selenium `WebDriver` methods such as `click()`. Once again, we use an abstract class that has dummy implemented method bodies. Each individual concrete class will only implement the required methods.

We start with the `abstract` class with the unimplemented methods:

```
public abstract class AActionKeyword extends TestBase {

    public AActionKeyword() throws IOException, FilloException {
        super();
    }
    public void clickElement(String x) {
    }
    public WebDriver navigate(WebDriver driver2, String url) {
        return null;
    }
    public WebDriver openBrowser(List<String> browsers) {
        return null;
    }
    public void sendKeys(WebElement elem, String textFill) {
    }

    public void selectValue() {
    }

}
```

Then comes the `Concrete` class with the implementations. Here, we open a Firefox driver, but we can parameterize this method to accept a string argument for the browser name:

```
public class OpenBrowser extends AActionKeyword {
        public WebDriver openBrowser() { WebDriver driver = new
ChromeDriver(); return driver; }
}
```

The following code is for the sample driver script with a single execution. This script will simply open a Firefox browser for now:

```
public class MainClass {
  public static void main(String[] args) throws IOException, FilloException
{
    WebDriver driver = null;
    List<String> browsers = new ArrayList<String>();
    browsers.add("chrome");
    System.setProperty("webdriver.chrome.driver",
        "src/main/resources/chromedriver.exe");
    AActionKeyword actKeyword = new OpenBrowser();
    ACommand command = new NewBrowser(actKeyword);
    BrowserInvoker invoker = new BrowserInvoker(command);
    driver = invoker.open(browsers);

  }
}
```

A look at the project structure in Eclipse

Here is what our project structure looks like in Eclipse. We will be adding content to all of these packages:

- `org.packt.client` will have the client code that drives the framework
- `org.packt.command` will have the `ICommand` interface and the `ACommand` dummy class for the various commands such as `click` and `enterText`
- `org.packt.invokers` will have the code for all invokers
- `org.packt.listeners` will have listener code
- `org.packt.receivers` will have the `IActionKeyword` interface, `AActionKeyword` dummy class, and the various receivers
- `org.packt.testbase` will have the `testbase` code
- `org.packt.utilities` will have utility methods

The following screenshots shows the project structure in Eclipse, with all the packages displayed:

Introducing the TestNG framework

We will now start the part that every test automation developer should know, that is, the TestNG framework. TestNG is a framework similar to JUnit and has many useful annotations. In this chapter, we will be using the `@Test` annotation. First, we will install the TestNG plugin for Eclipse.

Installing the TestNG plugin for Eclipse

Follow these steps to install the TestNG plugin for Eclipse:

1. Select **Help** | **Eclipse Marketplace**
2. Search for `TestNG Eclipse`
3. Install **TestNG for Eclipse**

The plugin's installation will take some time. Eventually, after the installation, we can right-click on the project and will see the various options that are available, as shown in the following screenshot:

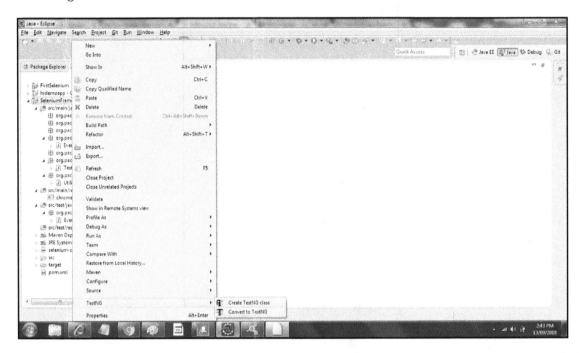

Click `Convert to TestNG`. On the screen that appears, click **Finish**. After a split second, you will see that a file with the name `testng.xml` has been created at the root location.

A look at testng.xml

This is what the `testng.xml` file looks like:

```xml
<?xml version="1.0" encoding="UTF-8"?>
<!DOCTYPE suite SYSTEM "http://testng.org/testng-1.0.dtd">
<suite name="Suite">
  <test name="ChromeTest" enabled="true">
    <classes>
      <class name="org.packt.client.Driver_Script">
            <parameter name="browserName" value="chrome"></parameter>
      </class>
  </classes>
  </test>
    <test name="IETest" enabled="true">
```

```
      <classes>
        <class name="org.packt.client.Driver_Script">
                <parameter name="browserName" value="ie"></parameter>
        </class>
  </classes>
  </test>
    <test name="FirefoxTest" enabled="true">
      <classes>
        <class name="org.packt.client.Driver_Script">
                <parameter name="browserName" value="firefox"></parameter>
        </class>
  </classes>
  </test>

  </suite> <!-- Suite -->
```

As you can see, the `Driver_Script` class in the `org.packt.client` package gets executed. Let's look at the contents of the `Driver_Script` class next.

The `Driver_Script` class has a switch case for all of the different keywords, such as `openBrowser`, and `Navigate`, as you can see in the following block of code:

```
@Test
@Parameters({ "browserName" })
public void Run_Framework(String browserName) {
  List<String> browsers = new ArrayList<String>();
  browsers.add(browserName);
  input_data = ExtractAllData.extractData(browserName);
  System.out.println("kya ha ?" + input_data.size());
  for (int i = 0; i < input_data.size(); i++) {
    String timestamp = new SimpleDateFormat("yyyyMMMddHHmmss")
        .format(new java.util.Date());

    System.out.println("Starting " + browserName + timestamp);
    actionKeyword = input_data.get(i).get("Keyword").toString();
    inputData = input_data.get(i).get("Data").toString();
    testcaseID = input_data.get(i).get("TestCaseID").toString();

    switch (actionKeyword) {
    case "openbrowser":
      try {
        actKeyword = new OpenBrowser();
      } catch (IOException e) {
        throw;
      } catch (FilloException e) {
        e.printStackTrace();
      }
```

```
          command = new NewBrowser(actKeyword);
          invoker = new BrowserInvoker(command);
          driver = invoker.open(browsers);
          break;
        case "navigate":
          try {
            actKeyword = new OpenURL();
          } catch (IOException e) {
            e.printStackTrace();
          } catch (FilloException e) {
            e.printStackTrace();
          }
          command = new OpenURLCommand(actKeyword);
          NavigatorInvoker invoker1 = new NavigatorInvoker(command);
          driver = invoker1.navigate(driver, inputData);
          break;
      }
    }
  }
}
```

If the `Framework.xlsx` file contains four test cases with two marked as `Y` in the `Execute_Flag` column, the `Browser` has the values `Chrome` and `Internet Explorer`, and if you execute `testng.xml` with this code, then one Chrome session will start, navigate to http://www.freecrm.com, and stay open. Similarly, an Internet Explorer session will open after that and navigate to the same URL and stay open. A new keyword called `closeBrowser` can be added if the browser windows need to be closed.

As you may have noticed from the timestamp, both of the timestamps will be different, indicating that each test case runs sequentially. But wait...if we have around 2,000 test cases in our test suite, won't the execution take a huge amount of time?

Incorporating Selenium Grid

Well, we have a solution to this. Here is where Selenium Grid comes to our rescue. We already set up a Selenium Grid Hub in Chapter 6, *The Actions Class and JavascriptExecutor*. Now, it's time to add a node. We will create batch files for creating the `hub` and `node`. The contents of the first batch file for the Hub is shown as follows. Name this file `seleniumgridhub.bat`:

```
java -jar selenium-server-standalone-3.14.0.jar -role hub
```

Double-click the `batch` file. The message, `Selenium Grid hub is up and running` is displayed

The contents of the second batch file for the node is shown in the following. Name this file `seleniumgridnode.bat`:

```
java -
Dwebdriver.chrome.driver=C:\Users\Bhagyashree\workspace\SeleniumFramework\s
rc\main\resources\chromedriver.exe -
Dwebdriver.ie.driver=C:\Users\Bhagyashree\workspace\SeleniumFramework\src\m
ain\resources\IEDriverServer.exe -
Dwebdriver.firefox.driver=C:\Users\Bhagyashree\workspace\SeleniumFramework\
src\main\resources\geckodriver.exe  -jar selenium-server-
standalone-3.14.0.jar -role node -hub http://localhost:4444/grid/register -
port 5555 -browser browserName=chrome,maxInstances=2,platform=WINDOWS -
browser browserName="internet explorer",maxInstances=2,platform=WINDOWS -
browser browserName=firefox,maxInstances=2,platform=WINDOWS
```

Note that the path shown in the following code has to be changed for the individual machine.

Double-click this file. A console window will appears, as follows:

```
C:\Windows\system32\cmd.exe                                          _ □ X
iver.chrome.driver=C:\Users\Bhagyashree\workspace\SeleniumFramework\src\main\res
ources\chromedriver.exe -Dwebdriver.ie.driver=C:\Users\Bhagyashree\workspace\Sel
eniumFramework\src\main\resources\IEDriverServer.exe -Dwebdriver.firefox.driver=
C:\Users\Bhagyashree\workspace\SeleniumFramework\src\main\resources\geckodriver.
exe  -jar selenium-server-standalone-3.14.0.jar -role node -hub http://localhost
:4444/grid/register -port 5555 -browser browserName=chrome,maxInstances=3,platfo
rm=WINDOWS -browser browserName="internet explorer",maxInstances=2,platform=WIND
OWS -browser browserName=firefox,maxInstances=2,platform=WINDOWS
16:04:55.320 INFO [GridLauncherV3.launch] - Selenium build info: version: '3.14.
0', revision: 'aaccce0'
16:04:55.331 INFO [GridLauncherV3$3.launch] - Launching a Selenium Grid node on
port 5555
2018-09-15 16:04:57.029:INFO::main: Logging initialized @2552ms to org.seleniumh
q.jetty9.util.log.StdErrLog
16:04:57.340 INFO [SeleniumServer.boot] - Selenium Server is up and running on p
ort 5555
16:04:57.340 INFO [GridLauncherV3$3.launch] - Selenium Grid node is up and ready
 to register to the hub
16:04:58.084 INFO [SelfRegisteringRemote$1.run] - Starting auto registration thr
ead. Will try to register every 5000 ms.
16:04:58.084 INFO [SelfRegisteringRemote.registerToHub] - Registering the node t
o the hub: http://localhost:4444/grid/register
16:05:01.006 INFO [SelfRegisteringRemote.registerToHub] - The node is registered
 to the hub and ready to use
```

Once the `Selenium Grid node is up and ready to register to the hub` message is displayed, navigate to `http://localhost:4444/grid/console`.

The screen that's shown is as follows:

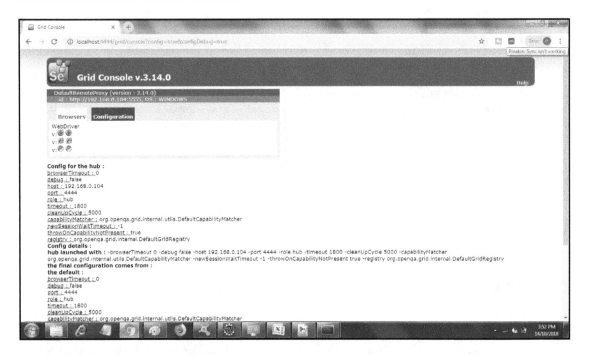

Let's understand why we got two Firefox, two Internet Explorer, and three Chrome instances. For this, we need to understand the script in the node batch file.

The content here is an excerpt from the `seleniumgridnode.bat` file:

```
Dwebdriver.chrome.driver=C:\Users\Bhagyashree\workspace\SeleniumFramework\s
rc\main\resources\chromedriver.exe
```

This line tells us that we should use the `webdriver.chrome.driver` argument to the actual `chromedriver.exe` file placed in the project resources folder:

```
-jar selenium-server-standalone-3.14.0.jar -role node -hub
http://localhost:4444/grid/register -port 5555
```

This registers the node to the hub on port `5555`:

```
-browser browserName=chrome,maxInstances=2,platform=WINDOWS
```

This creates two Chrome instances on the windows platform. This is the reason why we can see three Chrome instances in the node console. A similar concept applies to Firefox and IE.

Running tests in parallel

Now, we will turn our attention to the changes in the code. Let's start with testng.xml.

The suite tag needs to be changed, as shown here:

```
<suite name="Suite" parallel="tests">
```

This indicates that the test tags have to be executed in parallel. We are executing the same program, Driver_Script, three times, with different parameters for the browser name.

The next change has to be done at the place where we instantiate the various drivers. First of all, we create a DesiredCapabilities reference, static DesiredCapabilities capabilities;

The openBrowser (List<String> browserName) method should be changed; the code snippet of one if condition is shown as follows:

```
public WebDriver openBrowser(List<String> browserName)
    throws MalformedURLException {
  if (browserName.get(0).equalsIgnoreCase("chrome")) {
    log.info("Executing openBrowser");
    System.setProperty("webdriver.chrome.driver",
        System.getProperty("user.dir")
            + "\\src\\main\\resources\\chromedriver.exe");
    try {
      driver = new RemoteWebDriver(new URL(
          "http://localhost:4444/wd/hub"),
          DesiredCapabilities.chrome());
    } catch (MalformedURLException e) {
      e.printStackTrace();
    }
  } else if (browserName.get(0).equalsIgnoreCase("ie")) {
    System.setProperty("webdriver.ie.driver",
        System.getProperty("user.dir")
            + "\\src\\main\\resources\\IEDriverServer.exe");

    capabilities.setBrowserName("IE");
    try {
      driver = new RemoteWebDriver(new URL(
          "http://localhost:4444/wd/hub"),
          DesiredCapabilities.internetExplorer());
```

```
            } catch (MalformedURLException e) {
                e.printStackTrace();
            }
        } else if (browserName.get(0).equalsIgnoreCase("firefox")) {
            System.setProperty("webdriver.gecko.driver",
                System.getProperty("user.dir")
                    + "\\src\\main\\resources\\geckodriver.exe");
            driver = new RemoteWebDriver(
                new URL("http://localhost:4444/wd/hub"),
                DesiredCapabilities.firefox());
        }
        return driver;
    }
}
```

The `capabilities` object is instantiated to a Chrome capability. Next, we change the `ChromeDriver` constructor to `RemoteWebDriver`. The first argument to this is a URL object which accepts the HUB URL as the first argument and the `capabilities` object as the second argument.

Then, we execute the `testng.xml` by right clicking on **Run As TestNG suite**. Once the browser instances start opening, notice the changes that happen in the node console:

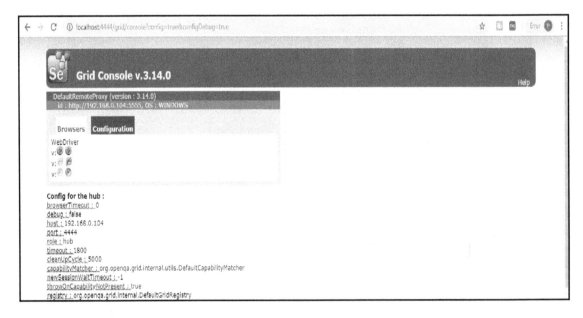

Notice how the Internet Explorer instances have become disabled. This indicates that we currently have one Chrome and one Internet Explorer session open.

Emailable report

After completion of the run, the TestNG framework generates a report called Emailable report, which is in HTML format and is available in the `test-output` folder in Eclipse.

The report shown here shows that there were two parameters, that is, Chrome and Internet Explorer:

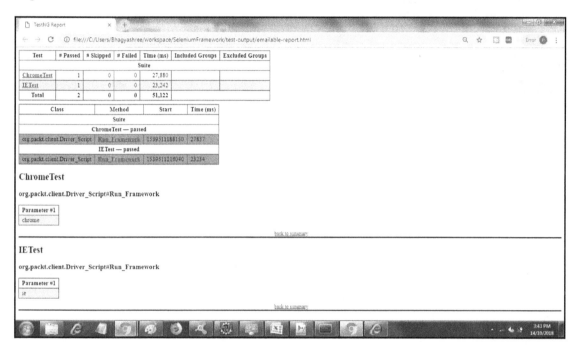

With this, we have completed this chapter on the command design pattern and the changes involved for incorporating Selenium Grid. In the next and final chapter, we will integrate the framework with logging, reporting, and Jenkins.

Summary

In this chapter, we took a sneak peek at one of the design patterns in Java, which is the command design pattern. We walked through the components of this behavioral pattern and developed the client, invoker, command, and receiver. We also took a look at how to create classes with empty implementations. Finally, we explored the setup of the Selenium Grid node and ran tests in parallel.

This was a very important chapter from a framework design standpoint. In the next chapter, we will touch upon the Data Provider and creating Excel reports. We will then look at TestNG listeners.

8
Hybrid Framework

We are in the final chapter, where we will explore two new concepts: TestNG Listeners and DataProviders. We will integrate our framework with reports and logging. In this chapter, we will look at the following concepts:

- Introducing `WebDriverManager`
- DataProviders in TestNG
- TestNG listeners and assertions
- Incorporating logging and reporting in the framework
- Steps to add keywords to the framework

Let's start the chapter with an introduction to a new concept: `WebDriverManager`.

Technical requirements

You will be required to have Ashot API in addition to software mentioned in Chapter 7, *The Command Pattern and Creating Components*.

The code files of this chapter can be found on GitHub:
https://github.com/PacktPublishing/Selenium-WebDriver-Quick-Start-Guide/tree/master/Chapter08

Check out the following video to see the code in action:
http://bit.ly/2O8KoXW

Introducing the WebDriverManager library

One constraint in Selenium automation is that we need to download the driver binary executable for Chrome, Firefox, and Internet Explorer. After downloading the executable files for Java, the absolute path of this executable file has to be set to JVM properties using System.setProperty. This was an overhead which has been removed with the introduction of a library called WebDriverManager.

 Remember to use JDK 1.8 with this library. It won't work with JDK 1.7.

With this library, there is no need to download the individual binaries for the different browsers. Earlier, we had to manage the versions of the binaries manually. With this library, this gets handled automatically.

How to use the WebDriverManager library

The WebDriverManager library can be used in three ways:

- As a Java dependency
- As a command-line interface
- As a server in itself

Let's look at each of these individually.

WebDriverManager as a Java dependency

One of the ways to use the WebDriverManager library is as a Java dependency. Add the dependency shown in the following code in the Dependencies section. This dependency can be found on GitHub:

```
<dependency>
    <groupId>io.github.bonigarcia</groupId>
    <artifactId>webdrivermanager</artifactId>
    <version>3.0.0</version>
    <scope>test</scope>
</dependency>
```

3.0.0 is the latest version at the time of writing.

Prior to `WebDriverManager`, we had to write the following in order to work with the browsers. Sample code for the Chrome browser is shown here:

```
private WebDriver driver = null;
  System.setProperty("webdriver.chrome.driver", "C:\\chromedriver.exe");
  driver = new ChromeDriver();
```

The same code can be written efficiently, as shown here:

```
private WebDriver driver = null;
WebDriverManager.chromedriver().setup();
driver = new ChromeDriver();
```

As we can seen, we don't require to set the binary path using `System.setProperty`. `WebDriverManager` handles it for us.

The following code fragment shows how to use `WebDriverManager` for Firefox:

```
private WebDriver driver=null;
WebDriverManager.firefoxdriver().setup();
driver = new FirefoxDriver();
```

The following code fragment shows how to use `WebDriverManager` for Internet Explorer:

```
private WebDriver driver=null;
WebDriverManager.iedriver().setup();
driver = new InternetExplorerDriver();
```

The sample code is shown here:

```
public class BrowserFactory {
    private WebDriver driver = null;
    public void openBrowser(String browser) {
        driver = null;
        WebDriverManager.chromedriver().setup();
        driver = new ChromeDriver();
    }
}
```

We will next see how to use `WebDriverManager` on the command line.

WebDriverManager as CLI

We can use `WebDriverManager` on the command line as well the the Java dependency. We do this simply by using the following Maven command:

```
mvn exec:java -Dexec.args="chrome"
```

On executing this command, the following logs get displayed in the console:

```
[INFO] Scanning for projects...
[INFO]
[INFO] ------------------------------------------------------------------------
[INFO] Building WebDriverManager 3.0.0
[INFO] ------------------------------------------------------------------------
[INFO]
[INFO] --- exec-maven-plugin:1.6.0:java (default-cli) @ webdrivermanager ---
[INFO] Using WebDriverManager to resolve chrome
[INFO] Reading https://chromedriver.storage.googleapis.com/ to seek chromedriver
[INFO] Latest version of chromedriver is 2.42
[INFO] Downloading https://chromedriver.storage.googleapis.com/2.42/chromedriver__win32.zip to
folder D:\projects\webdrivermanager
[INFO] Binary driver after extraction D:\projects\webdrivermanager\chromedriver.exe
[INFO] Resulting binary D:\projects\webdrivermanager\chromedriver.exe
```

WebDriverManagerServer

`WebDriverManager` can also be used as a server. There are two ways to use it as a server:

- Directly from the code and Maven
- Using `WebDriverManager` as a fat JAR

Let's look at both of these methods:

1. **Directly from the code and Maven**: The command to be used is: `mvn exec:java -Dexec.args="server <port>"`. If the second argument is omitted, the default port (`4041`) will be used: `C:\ Maven mvn exec:java -Dexec.args="server"`:

```
[INFO] Scanning for projects...
[INFO]
[INFO] ------------------------------------------------------------------
[INFO] Building WebDriverManager 3.0.0
[INFO] ------------------------------------------------------------------
[INFO]
[INFO] --- exec-maven-plugin:1.6.0:java (default-cli) @ webdrivermanager ---
[INFO] WebDriverManager server listening on port 4041
```

2. **Using `WebDriverManager` as a fat-jar:** We can also use WebDriverManager as a fat JAR. For instance: `java -jar webdrivermanager-3.0.0-fat.jar chrome`.

After executing this command, the following gets printed in the console:

```
[INFO] Using WebDriverManager to resolve chrome
[INFO] Reading https://chromedriver.storage.googleapis.com/ to seek chromedriver
[INFO] Latest version of chromedriver is 2.42
[INFO] Downloading https://chromedriver.storage.googleapis.com/2.42/chromedriver_win32.zip to folder D:\projects\webdrivermanager
[INFO] Resulting binary D:\projects\webdrivermanager\target\chromedriver.exe
```

When `WebDriverManager` gets up and running, HTTP requests can be done to resolve driver binary executables (`chromedriver`, `geckodriver`, and so on). For example, suppose that `WebDriverManager` is running the localhost and in the default port, navigates to the following URLs to get the latest versions of the driver executables:

- `http://localhost:4041/chromedriver`: The latest version of `chromedriver`
- `http://localhost:4041/firefoxdriver`: The latest version of `geckodriver`

- `http://localhost:4041/operadriver`: The latest version of `operadriver`
- `http://localhost:4041/phantomjs`: The latest version of `phantomjs` driver
- `http://localhost:4041/edgedriver`: The latest version of `MicrosoftWebDriver`
- `http://localhost:4041/iedriver`: The latest version of `IEDriverServer`

These requests can be initiated using a normal browser. The driver binary is then automatically downloaded by the browser as an attachment in the HTTP response when the response is sent back.

Advantages of WebDriverManager

Let's see what advantages the `WebDriverManager` offers:

- It checks the browser version installed on your machine (for example, Chrome, Firefox, Internet Explorer, and so on).
- It checks the version of the driver (for example, `chromedriver`, `geckodriver`). If the version is not known, it uses the latest version of the driver.
- It downloads the WebDriver binary executable if it is not present in the `WebDriverManager` cache in the Maven repository (`~/.m2/repository/webdriver` by default).
- It exports the appropriate WebDriver Java environment variables required by Selenium (not done when `WebDriverManager` is used from the the command-line interface or as a server).

Now that we know about the `WebDriverManager` library, we will look at what DataProviders are in TestNG.

DataProviders in TestNG

DataProviders in TestNG are used when we need to iterate over objects. These can be objects created from Java that include complex objects, objects from a property file or database, or simple string objects. Whenever a `dataProvider` attribute is specified in a `@Test` annotation, and a method with the annotation `@DataProvider` exists, the data from this method is passed into the `@Test` annotated method as a two-dimensional object array.

Let's look at a small example, which is enhanced from our earlier example, where we extracted data into a list of HashMaps. Here we go a step further and read the list of HashMaps in a two-dimensional object array. The following code demonstrates this concept. The code has been split into different sections for readability.

The following code fragment shows all the declarations for this program:

```
static Object[][] objectArray = {};
  static Object[][] noRecords = {};
  private static String tempParam;
  static Map<String, String> tempMap = new HashMap<String, String>();
  static List<HashMap> tempList = new ArrayList<HashMap>();
  private static List<HashMap> input_data;
  private static String browserName;
  static IActionKeyword actKeyword;
  private static Object actionKeyword;
  private static Object inputData;
  private static Object testcaseID;
  static ICommand command;
  static BrowserInvoker invoker;
  private static WebDriver driver;
```

The following code shows the @Test method with the @Parameters annotation for supplying the browser name and the @AfterTest method:

```
@Test()
  @Parameters("browserName")
  public static void fetchValues(String param) {
    tempParam = param;
  }

  @AfterTest
  public static void quitBrowser() {
driver.quit();
  }
```

The following code fragment shows the @Test method with the dependsOnMethods annotation and the data provider, Data1. This method has been made to depend on the fetchValues method using the dependsOnMethods attribute:

```
@Test(dataProvider = "Data1", dependsOnMethods = "fetchValues")
  public void getValues(String testCaseID, String teststepID,
      String action, String data, String keyword) {
    System.out.println("First Extract: " + testCaseID + ": " +
      teststepID + ": " + action + ": " + data + ": " + keyword);
    List<String> browsers = new ArrayList<String>();
```

```
    browsers.add(tempParam);
    String timestamp = new SimpleDateFormat("yyyyMMddHHmmss")
        .format(new java.util.Date());

    System.out.println("Starting " + browserName + timestamp);
    actionKeyword = keyword;
    inputData = data;
    testcaseID = testCaseID;
    System.out.println("Action Keyword: " + testCaseID);
    System.out.println("Action Data: " + data);

    switch (keyword) {
    case "openbrowser":
      try {
    actKeyword = new OpenBrowser();
    } catch (IOException e) {
    throw new IOException();
    } catch (FilloException e) {
    throw new FilloException("Error in query");
    }
      command = new NewBrowser(actKeyword);
      invoker = new BrowserInvoker(command);
      driver = invoker.open(browsers);
      break;
    case "navigate":
      try {
    actKeyword = new OpenURL();
    } catch (IOException e) {
    throw new IOException();
    } catch (FilloException e) {
    throw new IOException();
    }
      command = new OpenURLCommand(actKeyword);
      NavigatorInvoker invoker1 = new NavigatorInvoker(command);
     driver = invoker1.navigate(driver, data);
      break;
    }
  }
```

The following code shows the actual `DataProvider` method annotated with the `@DataProvider` annotation. This annotation is identified with the name `Data`, which can be referenced in the `@Test` annotation:

```
@DataProvider(name = "Data1")
 public static Object[][] exec() {
 String timestamp = new SimpleDateFormat("yyyyMMddHHmmss")
 .format(new java.util.Date());
```

```
System.out.println("Starting " + tempParam + timestamp);
Connection conn = null;
Fillo fillo = new Fillo();
Recordset rs = null, rs1 = null;
try {
conn = fillo.getConnection(System.getProperty("user.dir")
+ "\\src\\main\\resources\\Framework.xlsx");
String testConfig = "select * from TestConfig where Browser='"
+ tempParam + "'";
String allTestcases = "select * from TestCases";
 rs = conn.executeQuery(allTestcases);
String strQuery2 = "select * from TestCases where TestCaseID='";
rs = conn.executeQuery(testConfig);
for (int i = 0; i < rs.getCount(); i++) {
rs.next();
System.out.println(rs.getField("Browser"));
System.out.println(rs.getField("Execute_Flag"));
if (rs.getField("Execute_Flag").equalsIgnoreCase("y")) {
rs1 = conn.executeQuery(testcases
+ rs.getField("TestCaseID") + "'");

System.out.println("Rs1 count: " + rs1.getCount());
for (int j = 0; j < rs1.getCount(); j++) {
rs1.next();
tempMap = new HashMap<String, String>();
tempMap.put("TestCaseID", rs1.getField("TestCaseID"));
tempMap.put("TestStepID", rs1.getField("TestStepID"));
tempMap.put("Object", rs1.getField("Object"));
tempMap.put("Data", rs1.getField("Data"));
tempMap.put("Keyword", rs1.getField("Keyword"));
tempList.add((HashMap) tempMap);
}
}
}
objectArray = noRecords;
objectArray = new Object[tempList.size()][5];
for (int i = 0; i < tempList.size(); i++) {
objectArray[i][0] = tempList.get(i).get("TestCaseID");
objectArray[i][1] = tempList.get(i).get("TestStepID");
objectArray[i][2] = tempList.get(i).get("Object");
objectArray[i][3] = tempList.get(i).get("Data");
objectArray[i][4] = tempList.get(i).get("Keyword");
}
 } catch (FilloException e) {
return noRecords;
}
String timestamp1 = new SimpleDateFormat("yyyyMMMddHHmmss")
.format(new java.util.Date());
```

```
System.out.println("Ending " + tempParam + timestamp1);
return objectArray;
}
```

Having a separate DataProvider class

In the preceding code, we have used a data provider in the same class. A data provider with the @DataProvider annotation can be used in a different class. The changes that need to be done are shown here.

The fact is that the change has to be done only at one place and that is in the @Test annotation as shown here:

```
@Test(dataProvider = "Data1" , dataProviderClass =
ExternalProvider.class)
```

Next, we will see how to signal whenever any event happens. For this we use what are called TestNG Listeners.

Introducing TestNG listeners

We have already looked at WebDriver listeners. TestNG also provides listeners which are basically used to signal when a particular event happens, such as completion of a test method. TestNG listeners are used for logging and reporting purposes.

Different types of listeners

There are many different types of interfaces that allow us to modify TestNG behavior. Listed next are these interfaces:

- ITestListener: A listener used for listening on running tests.
- ISuiteListener: A listener used specifically for test suites.
- IReporter: This is an interface to be implemented if a report has to be generated.
- IMethodInterceptor: This is used to change the collection of methods that TestNG is going to execute.
- IInvokedMethodListener: This gets into action before and after a test method is invoked by TestNG.

- IHookable: If this interface is implemented by a class, then the run() method will be invoked instead of the @test method. The test method will be performed when the callBack() method of the IHookCallBack parameter is invoked.
- IAnnotationTransformer: The transform method will be invoked by TestNG to give you a chance to modify a TestNG annotation read from your test classes. You can change the values by calling any of the setter methods on the ITest interface.
- IAnnotationTransformer2: If any other test annotation needs to be modified apart from the @Test annotation, one can use this interface.

We will mainly be looking at the iTestListener and iSuiteListener interface implementations.

There are two ways in which TestNG listeners can be used:

- By extending the TestListenerAdapter class
- By implementing the corresponding interface, iTestListener

Let's understand both methods.

Implementing the ITestListener interface

For this, let's understand a very simple program. We create a very simple listener first.

We have implemented the ITestListener and four of the methods, which are:

- onTestStart
- onTestSuccess
- onTestFailure
- onTestSkipped

The following code shows the implementation of ITestListener:

```
public class SampleListener implements ITestListener {
    @Override
    public void onTestStart(ITestResult result) {
        System.out.println("The test that has started is: " +
result.getName());
    }
    @Override
    public void onTestSuccess(ITestResult result) {
```

```
    System.out.println("The test that has passed is: " + result.getName());
        }
    @Override
public void onTestFailure(ITestResult result) {
System.out.println("The test that has failed is: " + result.getName());
    }
    @Override
    public void onTestSkipped(ITestResult result) {
        System.out.println("The test that has skipped is: " +
        result.getName());
    }
    @Override
    public void onTestFailedButWithinSuccessPercentage(ITestResult
    result) {
        }
    @Override
    public void onStart(ITestContext context) {
        }
    @Override
    public void onFinish(ITestContext context) {
        }
    }
```

A question now arises: how do we use this listener to log messages and generate reports?

For this, we need to create a test class, as shown here:

```
@Listeners(org.packt.listeners.SampleListener.class)
public class ListenerClass {
 @Test
 public void checkListeners() {
     System.setProperty(
 "webdriver.chrome.driver",
System.setProperty("user.dir")+"\\src\\main\\resources\\chromedriver.exe");
     WebDriver driver = new ChromeDriver();
     driver.get("http://www.freecrm.com");
     System.out.println("Title is : " + driver.getTitle());
 }
}
```

When we right-click on this code and select **Run As | TestNG Test**, we get the following output:

```
The test that has started is: checkListeners
Title is : #1 Free CRM software in the cloud for sales and service
The test that has passed is: checkListeners
PASSED: checkListeners
```

Another way to use the listener class along with the test class is to use it as a tag in
TestNG XML. For this, we just need to add the following tag in the TestNG XML file:

```
<listener class-name="SampleListener"></listener>
```

Extending the TestListenerAdapter class

A second approach to implementing listeners is to extend the TestListenerAdapter
class.

The following code shows a sample listener class:

```
public class SampleListener1 extends TestListenerAdapter {
    public void onTestSuccess(ITestResult testResult) {
        System.out.println(testResult.getName() + " was a success");
    }
    public void onTestFailure(ITestResult testResult) {
        System.out.println(testResult.getName() + " was a failure");
    }
    public void onTestSkipped(ITestResult testResult) {
        System.out.println(testResult.getName() + " was skipped");
    }
}
```

The test class for the preceding listener will not change. The output is shown here:

**Title is : #1 Free CRM software in the cloud for sales and service
checkListeners was a success**

With this knowledge, let's have a look at the listener for our framework.

The UtilityListener class, shown in the following code, indicates that we have
implemented three interfaces:

```
public class UtilityListener implements ITestListener, ISuiteListener,
 IInvokedMethodListener {
    // This will execute before the Suite start
    @Override
    public void onStart(ISuite arg0) {
        Reporter.log("About to start Suite execution" + arg0.getName(),
        true);
    }
    // This will execute when suite has finished
    @Override
    public void onFinish(ISuite arg0) {
        Reporter.log("About to exit executing Suite " + arg0.getName(),
```

```
        true);
    }
    // This will execute before start of Test
    public void onStart(ITestContext arg0) {
        Reporter.log("About to begin executing Test " + arg0.getName(),
            true);
    }
    // This will execute, once the Test is finished
    public void onFinish(ITestContext arg0) {
        Reporter.log("Completed executing test " + arg0.getName(),
            true);
    }
    // This will execute once the test passes
    public void onTestSuccess(ITestResult arg0) {
    // This calls the printTestResults method
        printTestResults(arg0);
    }
    // This will execute only on the event of fail
    public void onTestFailure(ITestResult arg0) {
        printTestResults(arg0);
    }
    public void onTestStart(ITestResult arg0) {
        System.out.println("The execution of the main test starts");
    }
    // This will execute only if any of the main test skipped
    public void onTestSkipped(ITestResult arg0) {
        printTestResults(arg0);
    }
    public void onTestFailedButWithinSuccessPercentage(ITestResult
     arg0) {
    }
```

This printTestResults method, shown in the following code, will be executed in case our tests pass or fail:

```
    private void printTestResults(ITestResult result) {
        Reporter.log("Test Method resides in "
            + result.getTestClass().getName(), true);
        if (result.getParameters().length != 0) {
            String params = null;
            for (Object parameter : result.getParameters()) {
                params += parameter.toString() + ",";
            }
            Reporter.log(
             "Test Method had the following parameters : " + params,true);
            }
            String status = null;
            switch (result.getStatus()) {
```

```
        case ITestResult.SUCCESS:
        status = "Pass";
        break;
        case ITestResult.FAILURE:
            status = "Failed";
            break;
        case ITestResult.SKIP:
            status = "Skipped";
        }
        Reporter.log("Test Status: " + status, true);
    }
    public void beforeInvocation(IInvokedMethod arg0, ITestResult
     arg1) {
        String textMsg = "About to begin executing following method
         : " + returnMethodName(arg0.getTestMethod());
        Reporter.log(textMsg, true);
    }
    public void afterInvocation(IInvokedMethod arg0, ITestResult
     arg1) {
        String textMsg = "Completed executing following method : "
            + returnMethodName(arg0.getTestMethod());
        Reporter.log(textMsg, true);
    }
    private String returnMethodName(ITestNGMethod method) {
        return method.getRealClass().getSimpleName() + "."
            + method.getMethodName();
    }
}
```

Let's now explore another important part of the framework: assertions.

Introducing assertions

Testing in Selenium is incomplete if we don't have a way to verify our test results. The TestNG framework provides the `Assert` class, using which we can do validations and verifications in TestNG. We can use assertions to verify text on screen, do validations, and so on.

The most common methods in the `Assert` class are given here:

- `assertTrue(condition, message)`: Takes two arguments, which are `condition` and `message`. A check is done to see whether the Condition is `true` or not. If the Condition is evaluated as `false`, then it fails and the message gets printed. The message is optional.
- `assertFalse(condition, message)`: Takes two arguments, which are `condition` and `message`. A check is done to see whether the Condition is `false` or not. If the Condition is evaluated as `true`, then it fails and the message gets printed. The message is optional.
- `assertNull(object, message)`: Checks whether an object is null. If the object is not null, then it fails and the optional message is printed.
- `assertNotNull(object, message)`: Checks whether an object is not null. If the object is null, then it fails and the optional message is printed.
- `assertEquals(actual, expected, message)`: Checks whether the `actual` and `expected` are equal. If these are not equal, then it fails and the optional message is printed.
- `assertNotEquals(actual, expected, message)`: Checks whether the actual and `expected` are not equal. If these are equal, then it fails and the optional message is printed.

The following code is a simple example using assertions:

```
@Listeners(org.packt.listeners.SampleListener1.class)
public class ListenerClass {
 @Test
 public void checkListeners() {
     System.setProperty("webdriver.chrome.driver",
     System.getProperty("user.dir")
 + "\\src\\main\\resources\\chromedriver.exe");
     WebDriver driver = new ChromeDriver();
     driver.get("http://www.freecrm.com");
     System.out.println("Title is : " + driver.getTitle());
     Assert.assertEquals(driver.getTitle(),
         "#1 Free CRM software in the cloud for sales and service");
 }
}
```

Once this test is run after right-clicking on project and selecting **Run As TestNG Test**, the script navigates to the freecrm website and passes when it finds the correct title.

Testing for a negative scenario

You might wonder what happens when you give incorrect text in the `expected` parameter. Suppose we give the string `# Free CRM software in the cloud for sales and service`, then the `AssertEquals` check will fail and the output shown here will get printed to the console:

```
FAILED: checkListeners
java.lang.AssertionError: expected [# Free CRM software in the cloud for
sales and service] but found [#1 Free CRM software in the cloud for sales
and service]
```

Next, we will understand what are soft and hard asserts.

Two different types of asserts

There are two different types of assertions, hard assert and soft assert:

- **Hard assert**: When a hard assert is used and the test case fails, the statements following the assert will not get executed and the test execution will abort.
- **Soft assert**: When a soft is used and the test case fails, the statements following the assert will get executed and the test execution will pass.

The normal assert is a hard assert, which we have already seen. We will see how to use a soft assert next.

Using soft asserts

Unlike hard assertions, soft assertions allow us to continue execution on failure without aborting. For this, a class called `SoftAssert` is used. An object of this class has to be created in order to use it:

```
SoftAssert assertNew = new SoftAssert();
```

Now use this object to invoke the various methods. Let's have a look at this with a simple example:

```
public class SoftAssertExample {
    @Test
    public void checkListeners() {
        SoftAssert assertNew = new SoftAssert();
        System.setProperty("webdriver.chrome.driver",
        System.getProperty("user.dir")
```

```
              + "\\src\\main\\resources\\chromedriver.exe");
       WebDriver driver = new ChromeDriver();
       driver.get("http://www.freecrm.com");
       System.out.println("Title is : " + driver.getTitle());
          assertNew.assertEquals(driver.getTitle(),
              "# Free CRM software in the cloud for sales and
                  service");
       System.out.println("Assert has been executed");
   }
}
The output from the preceding code is given here:
```

If you check the output from this code, notice that even when the assertion fails, control passes to the statement following the assert, which in the preceding case is:

```
System.out.println("Assert has been executed");
```

```
   Title is : #1 Free CRM software in the cloud for sales and service
   Assert has been executed
    checkListeners was a success
    [Utils] Attempting to create
   C:\Users\Bhagyashree\workspace\SeleniumFramework\test-output\Default
   suite\Default test.xml
    [Utils] Directory C:\Users\Bhagyashree\workspace\SeleniumFramework\test-
   output\Default suite exists: true
    PASSED: checkListeners
   ===============================================
    Default test
    Tests run: 1, Failures: 0, Skips: 0
   ===============================================
```

Notice the `println` statement after the assert has been printed. Try the same code using a hard assert. You will notice that the `println` statement does not get printed and, moreover, the test fails.

Implementing logging and reporting in the framework

For implementing logging in the framework, define the `log` object in the `TestBase` class, as shown here:

```
protected static Logger log = null;
```

Mark the `AActionKeyword` class as shown here:

```
public class AActionKeyword extends TestBase implements IActionKeyword
```

Then initialize the `logger` object in the `OpenBrowser` constructor using the following code:

```
log = Logger.getLogger(OpenBrowser.class);
PropertyConfigurator.configure("log4j.properties");
```

With this, we can incorporate the `log4j` `logs` in the `OpenBrowser` class, as shown here:

```
public class OpenBrowser extends AActionKeyword {
    public OpenBrowser() throws IOException, FilloException {
        super();
        log = Logger.getLogger(OpenBrowser.class);
        PropertyConfigurator.configure("log4j.properties");
    }
    static RemoteWebDriver driver;
    static DesiredCapabilities capabilities;
    public WebDriver openBrowser(List<String> browserName) {
    if (browserName.get(0).equalsIgnoreCase("chrome")) {
        log.info("Executing openBrowser");
        System.setProperty("webdriver.chrome.driver",
System.getProperty("user.dir")
+ "\\src\\main\\resources\\chromedriver.exe");
        driver = new ChromeDriver();
    } else if (browserName.get(0).equalsIgnoreCase("ie")) {
        System.setProperty("webdriver.ie.driver",
System.getProperty("user.dir")
+ "\\src\\main\\resources\\IEDriverServer.exe");
        driver = new InternetExplorerDriver();
    } else if (browserName.get(0).equalsIgnoreCase("firefox")) {
        System.setProperty("webdriver.gecko.driver",
System.getProperty("user.dir")
+ "\\src\\main\\resources\\geckodriver.exe");
driver = new FirefoxDriver();
    }
    return driver;
  }
}
Output:
2018-09-23 12:56:04 INFO  OpenBrowser:31 - Executing openBrowser
```

Incorporating reporting in the framework

Apart from extent reports, we can create Excel reports from `testng-results.xml`. `testng-results.xml` gets generated in a test-output folder. We can make use of this XML to generate the Excel reports.

Creating a custom XL file from testng-results.xml

The following code shows how to generate the Excel report:

```
public class ReportGenerator {
 public static void main(String[] args) throws
ParserConfigurationException,
 SAXException, IOException {
     String destFile = "ReportGen.xls";
     String path = ReportGenerator.class.getClassLoader().getResource("./")
                   .getPath();
     path = path.replaceAll("target/classes", "test-output");
     File file = new File(path + "testng-results.xml");
     DocumentBuilderFactory docFactory = DocumentBuilderFactory
                                         .newInstance();
     DocumentBuilder docbuilder = docFactory.newDocumentBuilder();
     Document dcmt = docbuilder.parse(file);
     dcmt.getDocumentElement().normalize();
     XSSFWorkbook book = new XSSFWorkbook();
     NodeList tlist = dcmt.getElementsByTagName("test");
     for (int i = 0; i < tlist.getLength(); i++) {
         int rw = 0;
         Node tnode = tlist.item(i);
         String tname = ((Element) tnode).getAttribute("name");
         XSSFSheet sheet = book.createSheet(tname);
         NodeList classlist = ((Element) tnode)
                             .getElementsByTagName("class");
         for (int j = 0; j < classlist.getLength(); j++) {
             Node cnode = classlist.item(j);
             String cname = ((Element) cnode).getAttribute("name");
             NodeList tmethodList = ((Element) cnode)
                                 .getElementsByTagName("test-
              method");
             for (int k = 0; k < tmethodList.getLength(); k++) {
                 Node tmethodNode = tmethodList.item(k);
                 String tmethodname = ((Element) tmethodNode)
                                     .getAttribute("name");
                 String tmethodstatus = ((Element) tmethodNode)
                                     .getAttribute("status");
                 XSSFRow row1 = sheet.createRow(rw++);
```

```
                    XSSFCell cel1 = row1.createCell(0);
                    cel1.setCellValue(cname + "." + tmethodname);
                    XSSFCell cel2 = row1.createCell(1);
                    cel2.setCellValue(tmethodstatus);
                }
            }
        }
        FileOutputStream fstream = new FileOutputStream(path + "report/"
                                                            + destFile);
        book.write(fstream);
        fstream.close();
        System.out.println("Report Generated");
    }
}
```

Next, we will see the approach to be followed to add new keywords.

Adding keywords to the framework

We will see how the logic of the framework flows and this will give you an idea as to how to add new keywords to the framework:

1. The TestNG file will be created, as shown in the preceding code.
2. Decide whether the run will be sequential or parallel.
3. Design the ReadFillo class, as shown previously, for reading the framework sheet based on the browser parameter passed.
4. Based on the keyword fetched from each row of Excel, create the object of that keyword.

Next, we will look at another method for creating reports directly, using the Fillo API.

Creating reports using Fillo

During the test execution, reporting for each step is possible using a user-defined class, which can have all the fillo methods, such as insertRecord and updateRecord. insertRecord will insert a new record into the report file and updateRecord will update an existing record.

The following class provides the code for our class called `FilloReports.java`:

```
public class FilloReports {

  private Formatter x;

  public void createFile() {
    String timestamp1 = new SimpleDateFormat("yyyyMMMddHHmmss")
        .format(new java.util.Date());
    try {
      x = new Formatter("Results.xls" + timestamp1);
    } catch (FileNotFoundException e) {
        throw new FileNotFoundException();
    }
  }

  public void insertRecord(String testcaseID, String result) {
    String query = "insert into report (TestcaseID, Result) values ('"
+ testcaseID + "','" + result + "')";
  }

  public void updateRecord(String testcaseID, String result) {
    String query = "update report set result='" + result
        + "' where TestCaseID='" + testcaseID + "')";
  }

}
```

With this, we complete logging and reporting in the framework. Next, let's see how we can generate screenshots in Selenium.

Generating screenshots in Selenium

Generating screenshots for our test execution is an important part of the framework. It's as equally important as generating logs and reports.

The following code shows the traditional way of taking screenshots:

```
public class TakeScreenShot {
  public void takeScreenPrint(String[] args) {
    System.setProperty("webdriver.chrome.driver",
  System.getProperty("user.dir")
+ "\\src\\main\\resources\\chromedriver.exe");
    WebDriver driver = new ChromeDriver();
    driver.manage().window().maximize();
    driver.get("http://www.google.com");
```

```
File src = ((TakesScreenshot) driver).getScreenshotAs(OutputType.FILE);
try {
  FileUtils.copyFile(src, new File("C:/selenium/error.png"));
}
catch (IOException e) {
  throw new IOException();
}
}
}
```

This method is good only to take screenshots of visible areas of the screen. What if the page has a scroll bar and the full page is visible only if you scroll down?

This is where the AShot API comes to our rescue.

Using the Ashot API

The Ashot API provides the following features:

- Takes a screenshot of an individual `WebElement` on different platforms (such as desktop browsers, iOS simulator, mobile Safari)
- Decorates the screenshots
- Provides screenshot comparison

Taking screenshots of individual `WebElements` is also possible with AShot.

AShot takes a screenshot in three steps:

1. Captures a screenshot of the full page
2. Finds the element's size and coordinates
3. Adjusts the original screenshot by cropping

This way, AShot provides an image of the `WebElement`

To use AShot, add the dependency given here to `pom.xml`:

```
<dependency>
 <groupId>ru.yandex.qatools.ashot</groupId>
 <artifactId>ashot</artifactId>
 <version>1.1</version>
</dependency>
```

The following code helps to take a full-page screenshot:

```java
public class TakeScreenShot {

  public static void main(String[] args) {

    // Open Firefox
    System.setProperty("webdriver.chrome.driver",
  System.getProperty("user.dir")
  + "\\src\\main\\resources\\chromedriver.exe");

    WebDriver driver = new ChromeDriver();

    // Maximize the window
    driver.manage().window().maximize();

    // Pass the url
    driver.get("http://www.freecrm.com");

    Screenshot screenshot = new AShot().shootingStrategy(
        ShootingStrategies.viewportPasting(1000))
        .takeScreenshot(driver);
    try {
      ImageIO.write(screenshot.getImage(), "PNG",
          new File(System.getProperty("user.dir") + "//test.png"));
    } catch (IOException e) {
      throw new IOException();
    }
  }
}
```

The two lines mentioned here are sufficient to take a full-page screenshot:

```java
Screenshot screenshot = new AShot().shootingStrategy(
    ShootingStrategies.viewportPasting(1000))
    .takeScreenshot(driver);
try {
  ImageIO.write(screenshot.getImage(), "PNG",
      new File(System.getProperty("user.dir") + "//test.png"));
```

Next, we will see how to take screenshot of an individual `WebElement`.

The two lines are all that change for taking the screenshot of an individual `WebElement`.

Given here is the code to take a screenshot of an individual `WebElement`:

```
public class TakeScreenShot {

  public static void main(String[] args) {

    // Open Firefox
    System.setProperty("webdriver.chrome.driver",
        System.getProperty("user.dir")
            + "\\src\\main\\resources\\chromedriver.exe");

    WebDriver driver = new ChromeDriver();

    // Maximize the window
    driver.manage().window().maximize();

    // Pass the url
    driver.get("http://www.freecrm.com");

    // Take screenshot and store as a file format
    // File src = ((TakesScreenshot)
    // driver).getScreenshotAs(OutputType.FILE);
    WebElement element = driver.findElement(By.name("username"));
    Screenshot screenshot = new AShot().shootingStrategy(
        ShootingStrategies.viewportPasting(1000)).takeScreenshot(
        driver, element);
    try {
      ImageIO.write(screenshot.getImage(), "PNG",
          new File(System.getProperty("user.dir") + "//test.png"));
    } catch (IOException e) {
      // TODO Auto-generated catch block
      e.printStackTrace();
    }
  }

}
```

We will look at some new location techniques in Selenium WebDriver 3 next.

Some extra location techniques in Selenium WebDriver 3

It's about time to discuss the other location techniques in Selenium WebDriver 3

The ones that we will be discussing are:

- `ByIdOrName`: This locator can be used with `WebElements` that have both an `id` and `name`
- `ByChained`: This locator can be used to find `WebElements` using a series of other location techniques
- `ByAll`: This locator finds all `WebElements` that match any of the locators in sequence

The best way to explain these is by giving an example for each. For this, we will be creating a custom web page in HTML:

```html
<html>
  <body>
    <p>
        First Name:
        <input type=text id=fid/>
    </p>
    <p>
        Last Name:
        <input type=text name=lname/>
    </p>
  </body>
</html>
```

It's time to see the `ByIdOrName` locator.

Understanding ByIdOrName

The way `ByIdOrName` works is that the ID takes precedence. If a particular element is not found with the ID that is specified, the search is made based on a name. If the element is not found after the second search, it throws a `NoSuchElementException`.

Let's have a look at the code to handle this.

The following code uses an implicit wait of 30 seconds for each element on the page. First, it searches based on the ID and then on the name. Notice that we have to invoke the constructor in order to use this locator:

```
public class ByIdOrNameDemo {

  public static void main(String[] args) {
    System.setProperty("webdriver.chrome.driver",
  System.getProperty("user.dir")
          + "\\src\\main\\resources\\chromedriver.exe");

    WebDriver driver = new ChromeDriver();
    driver.manage().timeouts().implicitlyWait(30, TimeUnit.SECONDS);
    driver.get("C:\\Users\\Bhagyashree\\Desktop\\Documents\\myHTML.html");
    driver.findElement(new ByIdOrName("fid")).sendKeys("FirstTest");
    driver.findElement(new ByIdOrName("lname")).sendKeys("LastTest");
  }
}
```

Next we will have a look at `ByAll`.

ByAll locator

We have introduced an address text area that has a class, ID, and name defined and we will zero in on this element using the series of locators

The following code uses the `ByAll` constructor, which takes in a variable number of locators as arguments:

```
public static void main(String[] args) {
  System.setProperty("webdriver.chrome.driver",
System.getProperty("user.dir")
+ "\\src\\main\\resources\\chromedriver.exe");
  WebDriver driver = new ChromeDriver();
  driver.manage().timeouts().implicitlyWait(30, TimeUnit.SECONDS);
  driver.get("C:\\Users\\Bhagyashree\\Desktop\\Documents\\myHTML.html");
  driver.findElement(
      new ByAll(By.className("tarea"), By.id("tid"), By.name("tname")))
      .sendKeys("FirstTest");

  }
}
```

It's the turn of the `ByChained` locator.

ByChained Locator

With this locator, DOM traversal becomes extremely simple. `ByChained` traverses from the first locator in the variable argument list and continues until the last element locator in the list.

We have introduced one more test textbox, which is inside a span, and the span is inside a div. Let's see the final structure of the HTML document.

The following HTML code shows the structure of the sample page:

```
<html>
  <body>
    <p>
    First Name:
    <input type=text id=fid/>
    </p>
    <p>
    Last Name:
    <input type=text name=lname/>
    </p>
    <p>
    Address:
    <input type=textarea class="tarea" id="tid" name="tname"/>
    </p>
    <div class="div1">
      <span id="spn1">
          Test:
        <input type=text name=testname/>
      </span>
    </div>
  </body>
</html>
```

The following code walks through a chain of locators starting from the parent `div`, taking the `span` next, right into the final test textbox:

```
public class ByChainedDemo {
 public static void main(String[] args) {
 System.setProperty("webdriver.chrome.driver",
 System.getProperty("user.dir")
 + "\\src\\main\\resources\\chromedriver.exe");
 WebDriver driver = new ChromeDriver();
 driver.manage().timeouts().implicitlyWait(30, TimeUnit.SECONDS);
 driver.get("C:\\Users\\Bhagyashree\\Desktop\\Documents\\myHTML.html");
 driver.findElement(
 new ByChained(By.className("div1"), By.id("spn1"), By
```

```
    .name("testname"))).sendKeys("FirstTest");
    }
}
```

With this, we complete the three additional locator mechanisms.

We have explored Selenium 3 throughout this book. It's time to see what Selenium has in store for us in the future.

Welcome Selenium 4

Selenium 4 is going to change the history of Selenium with its new features. Let's see what the features are:

- Complies completely with W3C standardization
- For communicating with the browser, the JSON wire protocol will be compliant with W3C specifications
- New Selenium IDE available with Chrome and Firefox (currently in the beta stage)
- New command-line interface runner that is based on nodejs and supports parallel execution
- Improvements to Selenium Grid with the removal of thread safety bugs and improvements to Docker support
- Improvements to Selenium UI Grid

Many new features are being added and it's an interesting journey ahead.

Next steps

As a beginner to Selenium WebDriver, you will face many initial challenges. To overcome these, I recommend the following steps:

1. Be up to date with the Selenium documentation.
2. If you find the command pattern difficult to comprehend, create a simple driver script with an `ActionKeyword` class. Two classes should be more than enough.

3. Fillo is a great API but still has some limitations. If your company has a database that you can use, go ahead and implement the framework in a database. This will give you more power over joins and complex queries, which can help in retrieving information.

4. Read books on other frameworks, such as the Page Object Model.

5. Read Selenium blogs.

Summary

This last chapter has covered many important topics. We had a look at `WebDriverManager`, DataProviders, and so on. Apart from putting the framework pieces together, we had a look at how to create an Excel report from `testng-results.xml`. We had a look at some extra locators and, finally, we introduced Selenium 4.

I hope you enjoyed reading this book. This book has focused on Selenium concepts primarily and discussed various pitfalls that can be encountered.

I wish you an exciting journey with Selenium!

Other Books You May Enjoy

If you enjoyed this book, you may be interested in these other books by Packt:

Mastering Selenium WebDriver 3.0 - Second Edition
Mark Collin

ISBN: 978-1-78829-438-6

- Provide fast, useful feedback with screenshots
- Create extensible, well-composed page objects
- Utilize ChromeDriver and GeckoDriver in headless mode
- Leverage the full power of Advanced User Interactions APIs
- Use JavascriptExecutor to execute JavaScript snippets in the browser through Selenium
- Build user interaction into your test script using JavascriptExecutor
- Learn the basics of working with Appium

Selenium WebDriver 3 Practical Guide - Second Edition
Unmesh Gundecha

ISBN: 978-1-78899-976-2

- Understand what Selenium 3 is and how is has been improved than its predecessor
- Use different mobile and desktop browser platforms with Selenium 3
- Perform advanced actions, such as drag-and-drop and action builders on web page
- Learn to use Java 8 API and Selenium 3 together
- Explore remote WebDriver and discover how to use it
- Perform cross browser and distributed testing with Selenium Grid
- Use Actions API for performing various keyboard and mouse actions

Leave a review - let other readers know what you think

Please share your thoughts on this book with others by leaving a review on the site that you bought it from. If you purchased the book from Amazon, please leave us an honest review on this book's Amazon page. This is vital so that other potential readers can see and use your unbiased opinion to make purchasing decisions, we can understand what our customers think about our products, and our authors can see your feedback on the title that they have worked with Packt to create. It will only take a few minutes of your time, but is valuable to other potential customers, our authors, and Packt. Thank you!

Index